Wireless Sensor
Networks
A Cognitive Perspective

Adaptation in Wireless Communications

Series Editor: Mohamed Ibnkahla

Wireless Sensor Networks: A Cognitive Perspective
Mohamed Ibnkahla

Wireless Sensor
Networks
A Cognitive Perspective

Mohamed Ibnkahla

CRC Press
Taylor & Francis Group
Boca Raton London New York

CRC Press is an imprint of the
Taylor & Francis Group, an **Informa** business

CRC Press
Taylor & Francis Group
6000 Broken Sound Parkway NW, Suite 300
Boca Raton, FL 33487-2742

First issued in paperback 2017

Version Date: 20120809

ISBN 13: 978-1-138-07615-0 (pbk)
ISBN 13: 978-1-4398-5277-4 (hbk)

Visit the Taylor & Francis Web site at
http://www.taylorandfrancis.com

and the CRC Press Web site at
http://www.crcpress.com

Contents

Preface

Every day, wireless sensor networks are gaining in popularity and applications. They are now widely used in key areas such as smart homes and buildings, intelligent transportation, health care, public health, military, food safety, water quality, smart power grid, industrial processes, precision agriculture, security, environment, and similar applications. Various types of data with different rates and requirements are transmitted through these networks. Users are always in need of better coverage, connectivity, security, and energy efficiency while looking for miniaturized, low-cost, and autonomous devices. Due to these requirements, classical techniques used in this field will soon reach their limitations and will no longer fulfill users' requirements.

Cognitive communications is a new concept that emerged a few years ago and has proven its importance, especially in the field of cognitive radio. This concept has been generalized to cover all aspects of the wireless communications system design.

Wireless sensor networks represent an excellent area where cognition and intelligence can be easily developed and exploited not only to benefit the network efficiency and user requirements but also to create new needs and applications. This is because wireless sensor networks are, by nature, distributed systems where information can be made available about everything in the deployment area. Information may include user needs, user requirements, environment conditions, network conditions, node-level information (such as battery level, transmission range, processing capabilities, and position information), and so on. Security is also an important issue where the cognitive concept can play a key role.

In the past few years, research on cognitive approaches in telecommunications has been scattered over a large number of conference and journal papers. This book presents the state of the art of this field and proposes a unifying view of the different cognitive approaches and methodologies. This book will be a benchmark that sets the foundations of cognitive communications and opens a new era for research in this field.

Organization of the Book

Chapter 1 sets the basics of the cognitive concept, reviews the different approaches in this field, and presents a generic architecture for cognition in wireless sensor networks.

Chapters 2–5 target specific issues that need to be addressed through cognition, starting from cognitive radio and spectrum access (Chapter 2) to routing protocols (Chapter 5). Chapter 2 is devoted to cognitive radio in the physical layer and dynamic spectrum access using multiple-input/multiple-output systems and cooperative diversity techniques. Chapter 3 covers joint adaptation in the physical layer (adaptive modulation and power) and in the medium access layer (adaptive sleep). Chapter 4 discusses the performance of cross-layer design (which can be seen as a very basic level of cognition) in addressing quality of service routing in multihop networks. This chapter covers a number of multihop networks in addition to wireless sensor networks. We made this choice so that the reader can see the different requirements for each network type and how they have been addressed through cross-layer design. A typical cognitive routing protocol called cognitive diversity routing is presented in Chapter 5. The cognitive property is implemented through the knowledge of some parameters in the environment such as channel status and power levels, as well as user/application preferences.

After the reader has gone through the above specific cognitive features (Chapters 2–5), Chapter 6 implements the more general cognitive concept where a large number of parameters, requirements, utility measures, and end-to-end goals are involved. The chapter uses the concept of weighted cognitive maps to improve network lifetime through optimizing routing, medium access, and power control while fulfilling end-to-end goals. The methodologies and concepts presented in Chapter 6 are very general and can involve a variety of parameters and objectives related to all layers of the protocol stack.

Chapter 7 was initially planned to cover hardware implementation of cognitive architectures in a comprehensive way. However, hardware implementation is still not mature in the field of cognitive wireless sensor networks. Therefore, we instead cover the important issue of the implementation of GPS/INS-enabled wireless sensor networks. Real-time node position information is required in many wireless sensor network applications and communication protocols. Several cognitive approaches need node position information in order to be implementable in real-world applications. Chapter 7 describes the key steps for hardware implementation of wireless sensor networks equipped with real-time positioning devices that can work even during GPS outages.

Required Background

A general knowledge in telecommunications is required (for example, a basic undergraduate course in digital communications). A basic knowledge of wireless communications is a plus, but it is not required.

The book has been organized such that an expert in wireless communications can understand the chapters independently. However, non-experts are expected to read the chapters in the order they have been presented for a comprehensive and in-depth understanding of this field.

Acknowledgments

I am deeply indebted to many people who helped me throughout the multiple phases of this project. My PhD and MSc students as well as Dr. E. Bdira and Dr. A. Noureldin have contributed to the writing and simulation/experimental results of the chapters: PhD student G. Vijay (Chapter 1), PhD student A. Abu Alkheir (Chapter 2), MSc Student X. Zhao (Chapter 3), PhD student A. ElMougy (Chapters 4 and 6), MSc student Z. El-Jabi (Chapter 5), MSc student C. Tang (Chapter 7), Dr. E. Bdira (Chapters 3 and 5), and Dr. A. Noureldin (Chapter 7). I am really grateful to all of them.

I would like to thank all the organizations and companies that have given support to my research during the past 10 years. This includes the Natural Sciences and Engineering Research Council of Canada (NSERC), Ontario Centers of Excellence, Ontario Ministry of Natural Resources, Ontario Research Fund (ORF) Wisense Project, Lunaris Inc., ElectroMagneticWorks Inc., IBM Canada, and NSERC DIVA Network.

Book Features

- This book is the first of its kind in this area and covers all layers of the protocol stack.
- More than 250 references are included; each chapter has its own references.
- More than 130 figures and 20 tables are included.
- An introductory chapter covers the state of the art in this field.
- Each chapter starts with an in-depth survey of the state of the art of the corresponding topic.
- All chapters are written in a tutorial style, making them easy to understand.
- In-depth descriptions of the different algorithms and protocols are included.
- Step-by-step analysis of the different systems through extensive computer simulations and illustrations are provided.

Acronyms

AAC	adaptive admission control
ACA	admission control algorithm
ACF	autocorrelation function
ACK	acknowledgment
ACS	autocorrelation sequence
ADC	analog-to-digital converters
AF	application framework
AI	artificial intelligence
AM	adaptive modulation
AMI	adaptive modulation with idle mode
ANN	artificial neural network
AODV	ad hoc on-demand distance vector
APL	application layer
AS	adaptive sleep
ASAM	adaptive sleep with adaptive modulation
AWGN	additive white Gaussian noise
BER	bit-error rate
BWRC	Berkeley Wireless Research Center
CACP	contention-aware admission control protocol
CB	connectivity based
CCAR	cognitive channel-aware routing
CCMR	cost and collision minimizing routing
CCS	code composer studio
CDME	cognitive decision-making engine
CDP	cell density packet
CDR	cognitive diversity routing
CEP	circular error probable
CME	change monitoring engine
CN	cognitive network
CPLD	complex programmable logic device
CPN	cognitive packet network

CPU	central processing unit
CS	carrier sensing
CSI	channel state information
CR	cognitive radio
CSD	cyclic spectral density
CSL	cognitive specification language
CSMA	carrier-sense multiple access
CSMA-CA	carrier sense multiple access with collision avoidance
CSN	cognitive sensor network
CTS	clear to send
DARAM	dual-access random access memory
DE	data entity
DGRAM	delay guaranteed routing and MAC
DPRAM	dual port random access memory
DSA	dynamic spectrum access
DSDV	destination-sequence distance vector
DSN	destination sequence number
DSP	digital signal processing
DSR	dynamic source routing
EDD	expected disconnection degree
EECCR	energy-efficient m-coverage and n-connectivity routing
EEHC	energy-efficient hierarchical cluster-based routing
EEMP	energy-efficient multihop polling
ETX	expected transmission count
FBR	flood-based routing
FC	fusion center
FCC	Federal Communications Commission
FND	first sensor node death, first node death
FPGA	field-programmable gate array
FSM	finite state machine
GA	genetic algorithms
GAF	geographic adaptive fidelity
GEAR	geographic and energy-aware routing
GeRaF	geographic random forwarding
GM	Gaussian-Markov
GPIO	general purpose input/output

GPRMC	GPS recommended minimum sentence
GPS	global positioning system
GRISP	gateway relocation algorithm for improved safety and performance
GSM	global system for mobile communication
GVGrid	QoS routing for VANETs
GVLL	generic virtual link layer
GyTAR	improved greedy traffic-aware routing
HAL	hardware abstraction layer
I/O	input/output
IC	integrated circuit
iCAC	interference-based fair call admission control
IMU	inertial measurement unit
IQRouting	interference-aware QoS routing
ISM	industrial, scientific, and medical
ISR	interrupt service routine
IT	interference temperature
KB	knowledge base
KF	Kalman filter
KP	knowledge plane
LAN	local area network
LAR	location-aided routing
LBI	low battery input
LBO	low battery output
LEACH	low-energy adaptive clustering hierarchy
LET	link expiration time
LOS	line of sight
LPP	linear programming problem
LURP	local update-based routing protocol
LUT	look-up table
MAC	medium access control
MANET	mobile ad hoc network
MCDMA	micro-programming controlled direct memory access
MCU	micro controller unit
ME	management entity
MEMS	micro-electromechanical system

MHR	MAC header
MIMO	multiple input multiple output
MIMU	micro-miniature inertial measurement unit
MMSPEED	multispeed
MURU	multihop routing for urban VANET
NHC	nonholonomic constraints
NMEA	National Marine Electronics Association
NPN	negative-positive-negative
NWK	network layer
OFDMA	orthogonal frequency division multiple access
OODA	observe, orient, decide, and act
OSAL	operating system abstraction layer
PAN	percentage of alive nodes
PAN	previous access network
PAP	power adaptation policy
PDF	probability density function
PEGASIS	power-efficient gathering in sensor information system
PER	packet error rate
PHY	physical layer
PIC	programmable interface controller
PLL	phase-locked loop
PLR	packet loss ratio
PNF	probability of node failure
PSD	power spectral density
PU	primary user
PV	photovoltaic
QAM	quadrature amplitude modulation
QoS	quality of service
QRDS	QoS routing and distributed scheduling
RAC	routing and admission control
RF	radio frequency
RMS	root mean square
RNNRL	random neural networks with reinforcement learning
ROC	receiver operating characteristics
ROM	read-only memory
RRE	ratio of remaining energy

RREC	route discovery message
RREP	route reply
RREQ	route request
RRS	robust routing and scheduling
RSSI	received signal strength indicator
RTS	request to send
SAN	software adaptable network layer
SAP	service access point
SARAM	single-access random access memory
SDR	software-defined radio
SGPR	stable group-path routing
SINR	signal-to-interference plus noise ratio
SINS	strap-down inertial navigation system
SNR	signal to noise ratio
SP	smart packet
SPI	serial peripheral interface
SPIN	sensor protocol for information via negotiation
SPTF	Spectrum Policy Task Force
SSP	security service provider
SSR	self-selective routing
STBC	space-time block coding
STF	space-time-frequency
TARA	topology-aware resource adaptation
TDMA	time division multiple access
TM	traffic matrix
UART	universal asynchronous receiver/transmitter
V2I	vehicle to infrastructure
V2V	vehicle to vehicle
VANET	vehicular ad hoc network
VLSI	very-large-scale integration
VoIP	voice-over Internet protocol
WCETT	weighted cumulative expected transmission time
WCM	weighted cognitive maps
WLAN	wireless local area network
WMN	wireless mesh network
WSN	wireless sensor network
ZDO	ZigBee device object

1

Introduction to Cognitive Approaches in Wireless Sensor Networks

1.1 Introduction

Wireless sensor networks (WSNs) are constituted of small-size, lightweight, low-power nodes deployed in large numbers. They are used in a variety of applications, such as environment monitoring, health care, precision agriculture, security, food safety, water quality monitoring, intelligent transportation, smart grid communications, and so forth. Advancement in the field of low-power very-large-scale integration (VLSI) and embedded systems, together with the convergence of communication and computing technologies, has enabled the miniaturization of the sensing, processing, and communication devices. This has led to an expansion in the application domain for sensor networks. Researchers now believe that WSNs are the key enabling technology for ambient intelligence where a network of these tiny sensing devices would enable environment-aware, personalized, and adaptive computing based on real-time requirements of end users. However, communication in sensor networks is challenged by the limited availability of energy. The multiple, often conflicting, optimization objectives have always been a challenge to achieve because the restricted interactions among the layers of the protocol stack makes it difficult to cater to the varied goals of the network elements while simultaneously catering to the end objective of the network as a whole.

To address these problems, cross-layer design approaches have been proposed. This design paradigm allows information sharing among layers and enables the joint optimization of problems at different layers. However, including information from all layers leads to reduced modularity and increased adaptation loops, making the system more complex to handle. These limitations have been acknowledged by the research community, and more holistic approaches are being investigated.

In order that sensor networks become part of pervasive computing environments, they need to become proactive rather than reactive. They must have the ability to learn the changes in the environment, infer from past behavior about the best course of action, and be able to predict future behavior based on what best suits the application needs. In other words, we are talking of introducing cognitive behavior in sensor networks.

Cognition refers to the ability to be aware of the environment, learn from past actions, and use that information to make future decisions that benefit the network. Unlike intelligence that focuses only on decision mechanisms, cognition focuses on information from the environment [1]. Thus the ability to learn becomes a key differentiator between a cognitive network and a noncognitive one [2]. To illustrate the idea of introducing cognition in a wireless network, consider the example in Figure 1.1. S_1 and S_2 are the source nodes that are trying to route data to destination nodes D_1 and D_2. Out of the available relay nodes, it is determined that node R_5 has the lowest link outage probability to D_1 and D_2. Hence S_1 starts routing data through R_5. In the meantime, S_2 also starts routing a high traffic of data through R_5 (indicated by the solid paths). When multiple source nodes start routing their data through this node, the route through R_5 may get congested. But a cognitive network with learning capabilities will be able to identify and

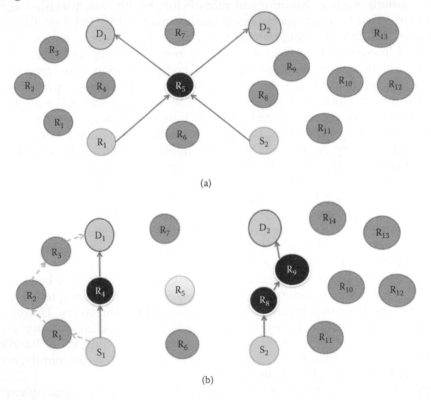

(a)

(b)

FIGURE 1.1
Traditional and cognitive routing in sensor networks. (a) Classical routing in a sensor network. (b) Cognitive routing in response to congestion.

predict the congestion at R_5 by observing the decrease in throughput at the source nodes, for example. Sharing this observation with all the nodes in the network, the cognitive network would be able to respond to congestion proactively by routing the data through a different path involving nodes R_4, R_8, and R_9 as shown in Figure 1.1(b). This helps to preserve nodes like R_5, thus maintaining network connectivity and providing reliable data transfer, which are especially important in sensor network applications. Based on the application, the network may be also able to choose between minimizing the number of hops (by choosing route $S_1{\rightarrow}R_4{\rightarrow}D_1$) and minimizing power, irrespective of the number of hops (by choosing route $S_1{\rightarrow}R_1{\rightarrow}R_2{\rightarrow}R_3{\rightarrow}D_1$), for example. Thus we can see that introducing cognition in a wireless network can be advantageous. Now let us go on to understand the objectives of introducing cognition in WSNs.

The following points summarize the objectives of incorporating cognition in sensor networks:

- Make the network aware of and dynamically adapt to application requirements and the environment in which it is deployed.

- Provide a holistic approach to enable the sensor network to achieve the end-to-end goals of the network; that is, gather information about the channel conditions from the physical layer, network status from network and MAC layers, and application requirements from the application layer, and then use memory of past actions and their outcomes in making informed decisions and optimizing the multiple objectives in the network.

- Enable the use of sensor networks in ambient intelligent environments.

Gathering information from all layers of the protocol stack will enable the sensor network to get a holistic view of the changes in the network: changes in application requirements, changes in the channel status at the physical layer, or the connectivity status of the network nodes. Let us look at the kind of information that is expected to be gathered from these network elements.

1.1.1 Application Layer Requirements

In a sensor network, end-user requirements may change over time or during a specific duration of time even if the deployed network is for a specific application. For example, in an environment monitoring application, coverage, connectivity, and high tolerance toward service disruptions are important aspects influencing the network lifetime. Here the normal function of the sensor network deployment is event monitoring. Hence the connectivity and coverage criteria play an important part in ensuring network lifetime maximization. After a certain duration of time, some nodes may die out, and this leads to a scenario where there is reduced data redundancy. There is now an increased need for reliable data transmission from the source node to the sink. Thus the reliability criterion gets added in addition to coverage and connectivity requirements. Moreover, if some nodes have cameras deployed, then there may be a demand for increased bandwidth when the end user wishes to turn on the camera modules at some chosen nodes.

In the case of a smart grid monitoring application, coverage, connectivity, real-time processing, bidirectional communication, and security are important requirements of this application. These requirements may change over time. For example, data transfer from photovoltaic (PV) panels does not need to take place at night. However, during daytime, the produced power and the stored power need to be communicated to the control center in quasi real-time. A similar example concerns the battery levels of electric vehicles, which do not need to be transmitted all the time. However, when electric vehicles request battery charging, their battery levels have to be transmitted to the grid's control center. During the charging process, the battery levels need to be transmitted as well. For the smart grid, data security requirements vary according to the deployed nodes and their respective roles.

The above scenarios explain how the application/end-user requirements may change over time for an application-specific deployment of a sensor network. Thus application requirements are representative of the end-to-end goals of the data flow in the network. Application demands should be given top priority during decision making and optimization in a cognitive sensor network.

1.1.2 Physical Layer Constraints and Requirements

The physical channel conditions such as path loss, signal-to-interference plus noise ratio (SINR), transmission power limitations based on remaining battery power at the node, and data rate constrain the physical layer and play a role in deciding whether the application requirements can be satisfactorily met or will have to be toned down. Hence the demands and constraints of the physical layer (PHY) are fully taken into account in cognitive decision-making while catering to the end-to-end network goals.

1.1.3 Network Status Sensors

The medium access control (MAC) and network (NWK) layers together provide information about the network status. While the network layer manages the routing scheme, connectivity, and role of the nodes (router/cluster head) in the network, the MAC layer handles node associations/disassociations, channel access control, enabling/disabling the radio, and beacon management. Security is handled by both layers. All this information from the NWK and MAC layers will be very useful in cognitive decision-making. For instance, when there is information available about the routing scheme from the network layer, channel conditions at the PHY, and application-layer requirements, a cognitive network may find that under the existing PHY conditions, the current routing scheme will not help achieve the demands of the end user. Hence, it may decide to instruct the NWK layer to adopt a different routing scheme because the current one is not able to cater to the network's end-to-end goals.

In order to achieve these objectives, learning the network conditions, having a memory of past actions, predicting future network conditions, and cognitive decision-making should be the core components of the system design.

Subsequent sections of the chapter are organized as follows: Section 1.2 presents the related work in the field of cognitive networks and cognitive sensor networks. A generic example of cognitive network architecture is presented in Section 1.3, followed by the conclusions in Section 1.4.

1.2 Related Work

Physical layer cognitive radio (CR) techniques [8] are the subject of Chapter 2. These techniques may be used in sensor networks and other networks. The reason for including CR in this book is that it represents an advanced and well-established physical layer awareness of the spectrum availability and use.

It is often believed that the layered architecture of the sensor network protocol stack hampers the network's ability to cater to multiple optimization objectives. Though cross-layer design has been popular, the interactions remain limited to a few layers. The network-wide performance goals are not accounted for. It provides reactive, memoryless adaptations of past outcomes for a given set of inputs. Cross-layer interactions lead to reduced architectural modularity, which in turn leads to increased instability and the high cost of maintaining the network. An overview of cross-layer design approaches is given in Chapter 4, which compares various cross-layer approaches and highlights their advantages and disadvantages.

A new approach is required to overcome the limitations of the existing design techniques. This technique could be a revolutionary one that is completely different from the existing ones or it could be an evolutionary one that builds on top of existing, proven-to-work techniques.

We explore the ideas proposed by Clark et al. [3], who proposed the concept of a knowledge plane (KP) for the wired networks. The KP is a construct that is different from the data and control planes of existing protocol stacks.

In the following sections, we look into the details of the KP and its functions, and how this concept was adapted into the wireless world. We also take a look at the various techniques employed toward realizing a cognitive sensor network (CSN).

1.2.1 Knowledge Plane and Cognitive Networks

The concept of knowledge plane (KP) was proposed by Clark et al. [3] in an effort to overcome the limitations and loopholes of the cross-layered design approach. The objective was to break the barriers of the layered structure and enable seamless communication across all the layers, as illustrated in Figure 1.2. KP was proposed to be a pervasive system based on knowledge rather than tasks, so that observations from different parts of the network could be correlated to make judgments in the presence of incomplete, inconsistent, or even conflicting information in dynamic environments. According to the authors in [3], the KP was expected to make decisions in the presence of partial or conflicting information, automate decisions, respond to emergency situations, and even foresee problems and proactively take corrective actions. The idea was to build a network that could assemble itself given high-level instructions, adapt itself to changes, reassemble if required, discover problems and fix them, or explain why it could not be fixed. Reasoning was expected to support the network's high-level goals and constraints and mediate between users or operators with conflicting goals and design constraints. The end-to-end goals of the entire network were kept in mind during optimizations, rather than those of a few interacting layers as in the cross-layer design approach. This required the system to have knowledge of the network and its actions and also have a learning mechanism to make decisions based on its past experiences. Hence, the tools of artificial intelligence (AI) and cognitive techniques of representation, learning, and reasoning were believed to be best suited to achieve the complex objectives of the KP as opposed to traditional algorithmic approaches.

Figure 1.2 represents an example of knowledge plane (KP) implementable on an IEEE 802.15.4/ZigBee stack. The strength of this concept lies in the information sharing across all layers through the KP.

The idea of the knowledge plane was adapted from the wired communication world to the wireless communication domain, and the concept of a cognitive network (CN) was proposed by Thomas et al. [4]. The CN paradigm spoke of the end-to-end scope of the network's goals, involving all elements within a data flow. It aimed at achieving these goals by breaking the layering up the network stack and communicating with nodes across the whole network.

The CN was defined to be a self-aware, self-organizing, and adaptive network capable of making intelligent adaptations based on:

- Observations of the network state made by individual elements
- Information sharing among nodes beyond the limitations of the layered protocol architecture
- Learning and reasoning before acting on its decisions to optimize network performance [5].

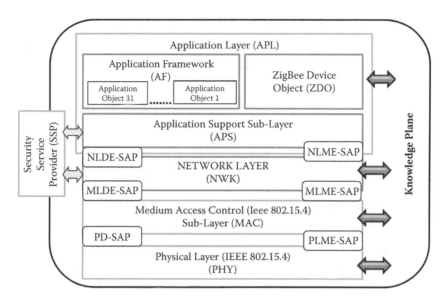

FIGURE 1.2
Knowledge plane in WSN protocol stack.

These CNs derive knowledge about the network performance from end users and applications, and have an end-to-end scope.

To implement the goals of the CN, the three-layer framework illustrated in Figure 1.3 was proposed by Thomas [6]. These layers consisted of the requirements layer, the cognition layer, and the software adaptable network (SAN) layer. The SAN was the architecture's interface with the physical world. Configurable network elements—such as directional antennas or cognitive radios (in which transmit power can be adaptive)—formed the action elements of the cognitive process. These were known as the modifiable elements. For each modifiable element, there was a one-on-one mapping with cognitive elements in the cognitive process in layer 2. These elements in the cognition layer helped to distribute the operation functionally and spatially. Network status sensors provided partial knowledge of the network to the cognitive elements. At the highest level of abstraction was the requirements layer, which transformed end-to-end objectives to goals for each cognitive element of the cognition layer through a cognitive specification language (CSL). The cognition layer was the central mechanism of this architecture. It learns about the system state, has knowledge about the current network goals, and decides on an appropriate response to observed network behavior. It makes use of a feedback loop in which past interactions with the environment guide current and future interactions. An "Observe, Orient, Decide, and Act" loop (or OODA loop [4]) was used as the feedback loop.

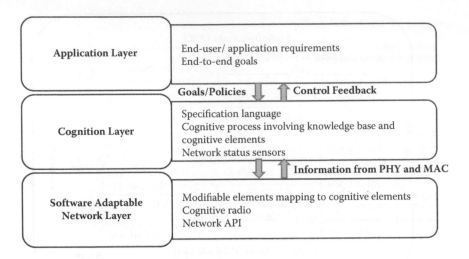

FIGURE 1.3
Example of cognitive network architecture.

These features enable learning and help the network converge on a solution faster than the network status changes. These ideas may be extended to the sensor network framework in order to make the network cognitive.

1.2.2 Cognitive Techniques Used in Sensor Networks

This section presents the latest trends in WSN research, which largely include cognitive approaches being adopted by researchers to improve the performance of WSNs.

1.2.2.1 Cognitive Radio in WSN

Cavalcanti et al. [7] present a conceptual design for CR-based WSNs and compare its performance with a standard ZigBee/802.15.4 WSN, both built on the standard model available in OPNET (ZigBee/802.15.4), operating in the 2.4 GHz band. In this experiment, Cavalcanti et al. [7] assumed the 802.15.4-based carrier-sense multiple access (CSMA) method in nonbeacon mode at the MAC layer and ZigBee-based protocols at the network layer (table-based mesh routing) and application layer for both the CR and 2.4 GHz modes. Transmit and receive antennas were both assumed to have the same unit gain. The receiver sensitivities were set at −85 dBm for the CR channel centered at 680 MHz, as well as for the first channel in the 2.4 GHz band. From these simulation results, the authors found that for the same transmit power, the maximum communication range in the CR channel is almost twice what is obtained in the 2.4 GHz channel. This increased range reduces the number of hops traveled per packet and hidden node problems, thus enhancing the efficiency of the multihop routing and the MAC. The overall application throughput was also found to be better in the CR mode.

Akan et al. [9] also talk about the main design principles, potential advantages, application areas, and network architectures of cognitive radio sensor networks. They explore the possibility of applying existing techniques for cognitive radio and WSN to such networks and identify the challenges in doing so.

1.2.2.2 Cognitive Schemes Using Neural Networks Models

Reznik and Pless [1] establish the feasibility of using distributed intelligence to embed cognition into sensor networks by studying the problem of signal change detection. They map artificial neural network (ANN) architecture to sensor networks and experimentally prove the advantage of this approach in terms of reduction in resource consumption: network bandwidth, processor power, and memory usage due to reduced connectivity and communication costs.

Youssef and Younis [10] propose a gateway relocation algorithm for improved safety and performance (GRISP), a neural network model to assess the safety of the gateway/sink node at various locations in a WSN environment trained using genetic algorithms. A "threat index" for each location visited, along with the snapshot captured, trains the neural network. This helps the neural network generate a "risk assessment factor" for making future safe relocation decisions.

1.2.2.3 Cognitive Sensor Networks

Shenai et al. [11] have presented a distributed WSN-based control system for intelligent and reliable operation of large power grids. Here sensor data (voltage and power factors) are reported to intelligent motes that communicate with nearby motes as well as a "control station" that can be mapped to the requirements layer. AUTOMAN, the software cognitive agents that motes and the control stations run, are said to have knowledge of local policies as well as awareness of the end-to-end operational requirements of the end application. Dynamic decisions are delivered and adequate information management is achieved by combining techniques of sensor coordination and intelligent data fusion. Thus, even under changing environments, dynamic reconfiguration is possible without grid downtime, and the system also ensures that the quality of service (QoS) requirements of the customer are always respected.

Boonma and Suzuki propose MONSOON [12], a biologically inspired framework to build cognitive WSN applications. This framework introspectively understands conflicting design objectives (data yield, data fidelity, power consumption), finds optimal tradeoffs with given constraints, and autonomously adapts to the dynamics of the network. It models an application as a decentralized group of software agents that collect sensor data from individual nodes and carry them to the base stations. From simulations, Boonma and Suzuki show that agents adapt to network dynamics by satisfying conflicting objectives under a given set of constraints and exhibit self-configuration, self-optimization, and self-healing properties.

1.2.2.4 Game-Theoretic Formulation of Energy Efficiency and Security in WSNs

Machado and Tekinay present a survey of game-theoretic formulation to the problems of energy efficiency and security in WSNs [13]. They found that distributed decision-making capabilities of WSNs and the selfish behavior of the individual nodes are exploited by the game-theoretic approach to optimize performance at the node level (conserving battery power) as well as the network level (maximizing the network utility, which is directly proportional to the number of sensors involved). They discuss the use of game-theoretic approaches in performing distributed cross-layer optimization by making use of the power control game at the physical layer and the rate allocation game at the application layer. In dealing with security issues in WSNs, Machado and Tekinay present work on game-theoretic models used to analyze situations where there are attacks by malicious nodes and outside intruders on WSNs. They also talk about the use of pursuit-evasion games in WSNs for model detection, tracking, and surveillance applications.

1.2.2.5 Cognitive and Self-Selective Routing

Gelenbe et al. [14] present cognitive packet network (CPN) and self-selective routing (SSR) algorithms that use different forms of learning as new approaches to achieving quality of service (QoS) routing in WSNs. CPN routing uses smart packets (SPs) for path discovery, along with random neural networks with reinforcement learning. It has the ability to adapt to varying traffic loads and is scalable for networks with flows to many destinations. The SSR technique, on the other hand, makes use of pheromone-based communication inspired by ants in a colony that communicate information about traversed paths to members of their kind. There is self-selection of routes at each node, which leads to additional overhead but provides the network with the ability to adapt to conditions where there are unexpected link failures or where the connections are unreliable. Both algorithms are capable of supporting fault tolerance to different degrees, and their protocol structure allows for different levels of efficiency in diverse application contexts.

We have provided an overview of several such techniques in [16]. Table 1.1 provides a brief summary of the literature reviewed in [16]. The following inferences can be drawn from Table 1.1:

TABLE 1.1

Comparison of Cognitive Techniques Applied to Sensor Networks

Technique	Goal Achieved by Cognition	Means of Achieving the Goals	Cognition Based on Knowledge/Learning/Reasoning/Context Awareness	Influence of Cognition on Network's End-to-End Goals
CR in WSN [7]	Increased communication range and application throughput	Implementing CR at PHY	Context awareness	No
ANN [2]	Reduced resource consumption by reducing connectivity and communication costs	Implementing distributed intelligence by mapping ANN to WSN architecture	Reasoning	Yes to a limited extent
GRISP [10]	Make safe relocation decisions for gateway node of WSN	Neural network model trained using genetic algorithms to assess safety of sink nodes	Learning	No
CSN [11]	Intelligent and reliable management and operation of large power grids, ensuring QoS requirements of end user are always respected.	Distributed sensors communicating with motes that have intelligent software agents called AUTOMAN, make system aware of end-user requirements and enable dynamic reconfiguration	Knowledge, context awareness	Yes
MONSOON [12]	Network exhibits self-configuration, self-optimization and self-healing properties by means of software agents	Decentralized group of software agents inspired by a biological framework that adapt to network dynamics by satisfying conflicting objectives under given set of constraints	Knowledge, context awareness	Yes to a good extent
Game Theoretic Formulation [13]	Network and node level performance improvement by game-theoretic formulation of problems of energy efficiency and security in WSNs.	Power control games at PHY, and rate control games at application layer for distributed cross-layer optimization; other game-theoretic models to analyze security issues in WSNs	May only be considered as an analysis tool; not cognition	No
Cognitive and Self-Selective Routing [14]	QoS routing in WSNs under diverse application contexts	Implementing different forms of learning such as: RNNRL and pheromone based techniques for cognitive routing	Reinforcement learning	No
Adaptive Modulation and Sleep Scheduling [15]	Better node lifetime in environment monitoring applications.	A cognition algorithm operating primarily on PHY and MAC layers, making use of feedback about channel conditions and modulation rate for adaptive sleep.	Learning from feedback	No

- The applications that were able to influence the network's end-to-end goals incorporated cognition based on learning, reasoning, or context awareness, or a combination of these.

- Cognitive techniques applied at the architecture/application level had a more pronounced network-wide impact than those that were introduced at specific layers of the protocol stack.

- Applications that used a distributed/decentralized approach to implementing cognitive decision-making were more successful in catering to the network deployment goals and changes in application requirements.

The results presented in [16] suggest the advantages of sharing information seamlessly across the layers of the network. Among all the approaches, the ones proposed by Shenai and Mukhopadhyay [11] and Boonma and Suzuki [12] are closest to the author's vision of applying cognition to sensor networks because they have a network-wide impact, are able to dynamically adapt to changing network conditions, and constantly track to the application requirements.

These cognitive techniques applied to sensor networks definitely promise improvements over the cross-layered approach, especially because they are based on knowledge and learning. However, these techniques do not explore the idea of a knowledge plane and do not have a definite framework that can be extended to all sensor network applications.

In the following, we present a generic architecture for cognitive nodes based on the concept of KP and a cognitive network framework [30]. When strategically deployed in a sensor network, these cognitive nodes will help to make the sensor network a cognitive one, and this concept can be applied to varied sensor network applications.

1.3 A Generic Architecture for Cognitive Wireless Sensor Networks

This section presents a generic cognitive node architecture that was originally developed by Vijay [30]. A number of cognitive nodes are deployed in the network. These nodes may or may not be equipped with sensors but are deployed with the purpose of improving the system's performance.

The objective of deploying a sensor network is said to be adequately achieved when the duration for which the network is functioning to deliver to its end-to-end goals is maximized. The end-to-end goals may involve every element of the network protocol stack—the application layer, network layer, MAC layer, and physical layer. They also include the goals of the data flow in the entire network. Hence, achieving the end-to-end goals of the network involves catering to multiple and often conflicting demands from the different layers. By using the cognitive techniques of learning and reasoning, together with intelligent optimization techniques, we will be able to better achieve the goals of a sensor network—thus enabling the maximization of application lifetime and enhancing end-user satisfaction.

The end-to-end goals of the network may include:

- Satisfying varying application requirements in a heterogeneous WSN environment
- Meeting physical layer requirements such as a fixed bit error rate (BER) or maintaining certain transmission power requirements
- Maintaining certain network coverage and connectivity requirements as nodes die out or change roles
- Scheduling frames and handling channel access as part of the MAC responsibilities

In order to achieve these goals, and without loss of generality, this approach brings modifications to the existing IEEE802.15.4/ZigBee protocol stack, but the concept can also be applied to other protocol stacks. To integrate learning, reasoning, decision making, and optimization into the existing protocol stack, the cognitive node architecture illustrated in Figure 1.4 is adopted.

Here the existing IEEE 802.15.4/ZigBee stack is extended to include the functions of learning, reasoning, decision making and optimization.

By introducing cognitive techniques at the node level, we will be able to introduce cognitive behavior in the entire sensor network by means of interaction among several such nodes interspersed in the network. The important features of this enhanced architecture are the cognitive decision-making engine (CDME), knowledge base (KB), change monitoring engine (CME), and the optimization engine. The expected capabilities of the cognitive nodes are as follows:

- At the application end, monitor any changes in user/application requirements and use the inputs in cognitive decision-making.
- Track information about the network status changes using inputs from the network status sensors.
- Store information about specific actions and their outcomes in the network in a knowledge base and look it up for predicting future behavior/cognitive decision-making.
- Get information from the physical layer about channel conditions before making decisions for the network. By monitoring spectrum-hole availability (see Chapter 2 of this book), the heterogeneous wireless environment may also be exploited in the unlicensed band.

We now describe the features of each component of the cognitive node.

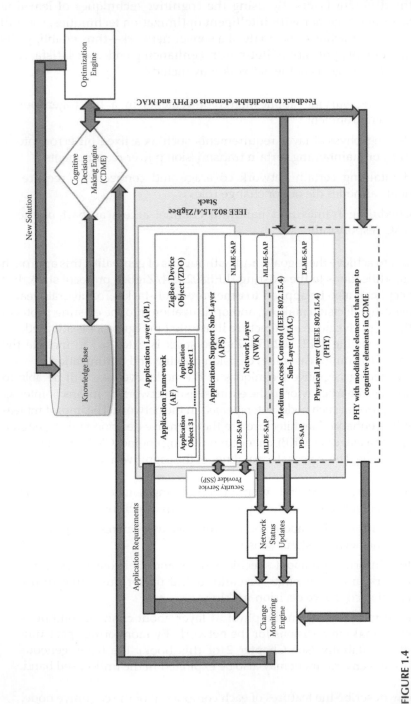

FIGURE 1.4

Proposed enhancements to the IEEE 802.15.4/ZigBee architecture to introduce cognition.

1.3.1 ZigBee Stack

IEEE 802.15.4-2006 [25] is the standard that defines the physical and medium access layers for radio frequency communication among WSN devices operating in the 2.4 GHz and 868/915 MHz license-free industrial, scientific, and medical (ISM) bands. ZigBee builds on top of the 802.15.4 standard and defines the network and application layers and a security service provider (SSP) [24]. Thus the ZigBee stack is comprised of IEEE 802.15.4-based PHY and MAC and ZigBee-based network and application layers. The different layers of the stack communicate with each other using service access points (SAPs) that are interfaced to the data entity (DE) or management entity (ME) services provided by a specific layer to the upper layers. Three types of devices are defined within a ZigBee network: a coordinator that starts and configures the network, a router that supports associations and forwards messages to other devices in the network, and an end device that communicates the sensed data to other devices in the ZigBee network. The ZigBee device object (ZDO) within the application layer (APL) defines these roles and is also responsible for initiating or responding to binding requests and securing relationships between network devices. Star, cluster tree, and mesh are the supported topologies, and the network layer supports all the functions related to starting a network, addressing, routing, and synchronization within the network. Manufacturer-defined application objects within the application framework (AF) implement the actual applications in accordance with the ZigBee-defined application descriptions. The IEEE 802.15.4 PHY and MAC along with ZigBee's network and application layers provide reliable data transfers over short ranges at very low power consumption, thus making it convenient to deploy sensor networks in a variety of applications.

1.3.2 Network Status Sensors

Network status sensors help track the changes in the MAC and NWK layers. Information such as the status of the nodes (end device/router/coordinator), their connectivity with neighboring nodes, information about new nodes that have joined the network or nodes that have temporarily been disassociated from the network, and the packet error rate are some of the inputs that can be provided as network status updates from these layers. Moreover, the IEEE 802.15.4/ZigBee stack has been set up in such a way that this information is inherently available and made use of in order to support the self-assembling feature of sensor networks. This information is used to help in cognitive decision-making.

1.3.3 Inputs from the Physical Layer

A lot of information can be gathered from the physical channel, such as the path loss, fading characteristics, shadowing properties, channel availability and capacity, interference level, and information about the remaining battery power in the nodes and the physical distance between nearest neighboring nodes, to name a few (see Chapters 2 and 3). This information could be useful in making decisions, such as estimating the transmit power required by the nodes to reach information from source to sink, choosing the best routing path based on channel conditions, energy consumption, and application QoS requirements such as reliability, throughput, and so forth. Many advanced protocols in sensor networks make use of such information. However, the cognitive nodes have an edge over the noncognitive networks because they provide a feedback path from the cognitive decision-making engine to the physical and MAC layers. The cognitive engine, based on its prediction of future behavior of the network, sends a feedback to these layers regarding changes that can be made to save power or modify some parameters to meet end-user requirements in future. The feedback could contain information regarding changes in transmission power or if the device can switch to a different mode (passive/active) of operation, for instance; it may also instruct the physical layer to change the modulation and coding schemes, and so on. Thus the feedback loop helps prepare the network to adapt to the predicted changes.

1.3.4 Change Monitoring Engine

The CME gathers updates from all the layers of the ZigBee protocol stack and provides it to the CDME, thus helping it "learn" the changes in the node's neighborhood. The frequency at which these updates are received can be time slotted and sent to the cognitive engine so that any changes in decisions can be made well in time so that the network lifetime is maximized and its performance unaffected.

1.3.5 Knowledge Base

The knowledge base (KB) is the storehouse of information in the cognitive node. It contains preprogrammed inputs based on the end-user requirements—that is, the target application that the sensor network is deployed for, its environment, and conditions at the physical layer (propagation environment, expected level of interference, and so forth). It gets updated with better and more precise information as the learning and optimization processes are activated. This way, the KB remains updated with the best decisions that can be made. The KB needs to store information about the node inputs, the associated decisions made during specific scenarios, and the impact of these decisions on the network, as it is closely associated with the decision-making process and aids the same. This means that the KB must have memory elements in it. In order to implement memory in its simplest form, the idea proposed here is to store previous decisions in the form of a look-up table (LUT) in the KB. Thus the KB would serve as a repository of how well the network had performed in a given scenario under the influence of a certain set of parameters. These inputs would help the CDME make proactive decisions under varying network conditions, thus contributing toward cognition in the network.

1.3.6 Cognitive Decision-Making Engine

The CDME is the heart of the cognitive process. It makes use of the information available from the network status sensors, physical layer, and application layer; combines it with previous information available in the knowledge base; and converts the available information to useful knowledge. Figure 1.5 represents the logical inputs that the cognitive decision-making engine of a cognitive node would require in order to make reliable predictions on node/network behavior and successfully achieve the network's end-to-end goals.

The CDME would have to associate itself with some form of learning and implement a decision-making mechanism to help in the cognitive process. Machine learning techniques are often used to develop algorithms that allow users to evolve behavioral patterns based on accumulated data. Forster [26] provides a valuable survey of several machine-learning techniques applied to wireless ad hoc and sensor networks.

A learning technique [23,26–29] will be needed for learning parameters from the physical, MAC, and network layers and also the end-application requirements. The learning process is activated once the network is functional. A possible pool of actions will be available based on a set of rules/observations.

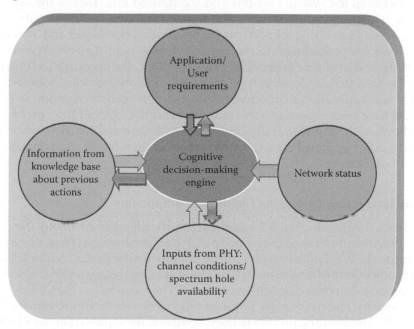

FIGURE 1.5
Inputs to the CDME.

To assist with the decision-making process, case-based decision theory can be used [21]. Wang et al. [22] have made use of case-based reasoning to solve the problem of clustering in sensor networks. However, our goal is to make the CDME work to achieve the network's end-to-end goals, tracking to the changes in the network's environment and end-user requirements.

The CDME can be implemented using a digital design technique: a Mealy state machine. In this state machine, the outputs depend not only on the current inputs but also on the outputs from the previous state. The states can be encoded, and the most significant bit could represent changes in the application requirement. The changes can be tracked in subsequent states, and further decisions on how the network should react could be made based on inputs in subsequent states from the other layers—PHY, MAC, and NWK in the same sequence (order of priority). Further possibility of splitting the state machine into two interacting subparts can also be explored. One state machine could be used for tracking local processes within the nodes and the other for tracking global changes due to interacting nodes.

1.3.7 Optimization Engine

We now explore the possibility of the combined use of learning and optimization processes to the advantage of the network. Feedback from the learning system can augment the optimization routines through comparisons between the system's actions and the desired outcomes of the optimization. The learning system can aid decision making in time-constrained situations by providing a known working solution developed and stored in the KB in the past, closest to the current system requirements. When sufficient time is available to provide better solutions, the optimization process may be activated to develop new solutions or evolve existing solutions to better suit application requirements. For example, genetic algorithms (GA) [18–20] have been chosen to solve multiple objective optimization problems. Furthermore, a local search algorithm (for example, a gradient descent-based search) may be applied after the global optimization has been performed to improve efficiency.

1.3.8 Interaction among the Cognitive Components

The block diagram in Figure 1.6 summarizes the interactions among the various components of the cognitive node. All the elements apart from those that belong to the WSN protocol stack can be imagined to be part of the knowledge plane proposed by Clark et al. [3]. They help achieve the goals that Clark et al. had proposed in their description of the KP. They also follow the OODA loop suggested by Thomas et al [4]. Local and global observations are gathered by the CME. The learning process in the CDME and the inputs from the KB help in orienting the nodes to cater to the network goals. The decision-making engine makes proactive decisions based on the knowledge it infers from the

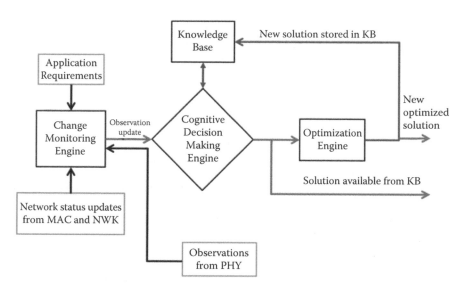

FIGURE 1.6
Simplified block diagram illustrating the interactions among the various blocks of the enhanced protocol stack.

CME and KB inputs. The output from the CDME may be further optimized by the optimization engine based on time and resource availability.

The outputs from the CDME (both optimized and unoptimized) are fed back to the PHY and MAC layers to orient them to the network conditions and application requirements, thus completing the OODA loop.

To summarize, the following features are needed for the cognitive node:

- Implement a knowledge base that can store useful information about specific actions in the network and read it back to help in making cognitive decisions for the network.

- Implement techniques such as reinforcement learning that will help in predicting future behavior based on previous actions and/or observations.

- The network status sensors must have associated processes that can identify useful information, discard irrelevant information, and communicate important information that helps to optimize network connectivity.

- Identify what kind of network status information will be useful in cognitive decision-making and extending network lifetime.

Once such a node is developed, to quantify the level of cognition achievable by the network where several cognitive nodes are deployed, the nine-level scale shown in Table 1.2 may be used. These levels of cognition were

TABLE 1.2

Levels of Cognition Achievable by a Cognitive Node

Level 0	*Preprogrammed:* Software radio
Level 1	*Goal Driven:* Makes radio/networking decisions according to goal; requires environment awareness
Level 2	*Context Awareness:* Knowledge of what the user is trying to do (distributed or central intelligence)
Level 3	*Radio/Network Aware:* Knowledge of radio and network components.
Level 4	*Capable of Planning:* Analyze situation (Level 2 and 3) to determine goals; follows a prescribed plan
Level 5	*Conducts Negotiations:* Settle on a plan with another radio, network, or processing entity
Level 6	*Learns Environment:* Autonomously determines structure of environment
Level 7	*Adapts Plans:* Generates new goals
Level 8	*Adapts protocols:* Proposes and negotiates new protocols

first discussed by Mitola and Maguire [18]. According to this scale, level 8 denotes the highest level of cognition the network can achieve, in which the node becomes capable of proposing and negotiating new protocols. The goal is to try and implement the highest level of cognition.

1.4 Conclusion

This chapter sets the foundations of cognitive wireless sensor networks and their architectures. We discussed several approaches in implementing cognition in WSNs. It should be noted that a complete implementation of the cognitive WSN (for example, including the nine levels of cognition) does not exist in the literature yet. What exist are attempts to implement only some aspects of the cognitive concept.

The remaining chapters of the book give examples of implementations of the above ideas, targeting a number of cognitive features. Chapter 2 targets cognitive radio networks (physical layer cognition) based on spectrum awareness and opportunistic spectrum access. Chapter 3 is devoted to adaptive modulation, adaptive power, and adaptive medium access. Chapter 4 presents cross-layer approaches in wireless sensor and ad hoc networks. Chapter 5 covers cognitive diversity routing (mainly focusing on physical and network layers). Chapter 6 employs cognitive engines that are aware of their environment and where several parameters are optimized while fulfilling end-to-end goals. Finally, knowing that several cognitive algorithms and protocols are based on node-location information, Chapter 7 investigates hardware implementation of location-based WSN nodes. In particular, the chapter emphasizes combined GPS/INS integration with ZigBee-based WSN nodes.

References

1. L. Reznik and G. Von Pless, "Neural networks for cognitive sensor networks," *IEEE International Joint Conference on Neural Network*, IJCNN 2008, June 2008, pp. 1235–1241.
2. D. H. Friend, "Cognitive networks: Foundation to applications," PhD diss., Electrical and Computer Engineering, Virginia Polytechnic and State University, Blacksburg, Va., March 6, 2009.
3. D. D. Clark, C. Partrige, J.C. Ramming, and J.T. Wroclawski, "A knowledge plane for the Internet," *Proceedings of the SIGCOMM* 2003, Karlsruhe, Germany, August 2003, pp. 3–10.
4. R. W. Thomas, L.A. DaSilva, and A.B. MacKenzie, "Cognitive networks," *2005 First IEEE International Symposium on New Frontiers in Dynamic Spectrum Access Networks*, DySPAN 2005, pp. 352–360.
5. R. W. Thomas, D.H. Friend, L.A. DaSilva, and A.B. MacKenzie, "Cognitive networks: Adaptation and learning to achieve end-to-end performance objectives," *IEEE Communication* 44, no. 12, December 2006, pp. 51–57.
6. R. W. Thomas, *Cognitive Networks*, PhD diss., Computer Engineering, Virginia Polytechnic and State University, Blacksburg, Va., June 15, 2007.
7. D. Cavalcanti, S. Das, J. Wang, and K. Challapali, "Cognitive radio based wireless sensor networks," *Proceedings of the 17th International Conference on Computer Communication and Networks*, 2008, ICCCN '08, pp. 1–6.
8. J. Mitola III and G.Q. Maguire, "Cognitive radio: Making software radios more personal," *IEEE Personal Communication* 6, no. 4, August 1999, pp. 13–18.
9. O.B. Akan, O. Karli, and O. Ergul, "Cognitive radio sensor networks," *IEEE Network* 23, no. 4, July–August 2009, pp. 34–40.
10. W. Youssef and M. Younis, "A cognitive scheme for gateway protection in wireless sensor network," *Applied Intelligence Journal* 29, no. 3, 2008, pp. 216–227.
11. K. Shenai and S. Mukhopadhyay, "Cognitive sensor networks," *IEEE 26th International Conference on Microelectronics (MIEL)*, May 2008, pp. 315–320.
12. P. Boonma and J. Suzuki, "Exploring self-star properties in cognitive sensor networking," *Proceedings of IEEE/SCS International Symposium on Performance Evaluation of Computer and Telecommunication Systems (SPECTS)*, Edinburgh, June 2008, pp. 36–43.
13. R. Machado and S. Tekinay, "A survey of game-theoretic approaches in wireless sensor networks," *International Journal of Computer and Telecommunications Networking* 52, no. 16, November 2008, pp. 3047–3061.
14. E. Gelenbe, P. Liu, B.K. Szymanski, M. Lisee, and K. Wasilewski. "Cognitive and self selective routing for sensor networks," *Journal of Computational Management Science*, August 14, 2009. Berlin-Heidelberg: Springer, vol. 6, pp. 1–22. Available online: www.springerlink.com/index/j351352425w41v68.pdf.
15. E. Bdira and M. Ibnkahla, "Performance modeling of cognitive wireless sensor networks applied to environmental protection," *Proceedings of the IEEE GLOBECOM'09*, Honolulu, Hawaii, 2009.
16. G. Vijay, E. Bdira, and M. Ibnkahla, "Cognition in wireless sensor networks: A perspective," Invited paper, *IEEE Sensors Journal* 11, no. 3, March 2011.

17. I. Dietrich and F. Dressler, "On the lifetime of wireless sensor networks," *ACM Transactions on Sensor Networks* 5, no. 1, February 2009, pp. 1–39.
18. D.B. Jourdan and O.L. de Weck, "Layout optimization for a wireless sensor network using a multi-objective genetic algorithm," *Vehicular Technology Conference, 2004. VTC 2004 Spring. 2004 IEEE 59th*, 2004, vol. 5, pp. 2466–2470.
19. H. Li, Y. Ding, Z. Zhang, and H. Zhang, "Node power management design in wireless sensor networks based on genetic algorithm," in *Networking and Digital Society (ICNDS)*, 2010 second international conference, 2010, pp. 191–194.
20. A.P. Bhondekar, R. Vig, M.L. Singla, C. Ghanshyam, and P. Kapur, "Genetic algorithm based node placement methodology for wireless sensor networks," *Proceedings of the International Multiconference of Engineers and Computer Scientists*, 2009, vol. 1, IMECS 2009, March 18–20, 2009, Hong Kong.
21. I. Gilboa and D. Schmeidler, "Case-based decision theory," *Quarterly Journal of Economics* 110, August 1995, pp. 605–639.
22. Y. Wang, K.W. Baek, K.T. Kim, H.Y. Youn, and H.S. Lee. "Clustering with case-based reasoning for wireless sensor network," in *Proceedings of the 2008 International Conference on Advanced Infocommunication*, Shenzen, China: ACM, NY, Article no. 24, 2008
23. S. Hussain and A.W. Matin, "Energy efficient hierarchical cluster-based routing for wireless sensor networks," Jodrey School of Computer Science, Acadia University, Wolfville, Nova Scotia Technical Report, pp. 1–33, 2005.
24. P. Kinney, "ZigBee technology: Wireless control that simply works," *ZigBee Alliance*, October 2003. Online. Available: http://www.zigbee.org/en/resources/
25. "IEEE standard for information technology: Telecommunications and Information exchange between systems, local and metropolitan area networks: Specific requirements Part 15.4: Wireless medium access control (MAC) and physical layer (PHY) specifications for low-rate wireless personal area networks (WPANs)," IEEE Std 802.15.4-2006 (Revision of IEEE Std 802.15.4-2003), pp. 0_1–305, 2006.
26. A. Forster, "Machine learning techniques applied to wireless ad-hoc networks: Guide and survey," in *Proceedings of the Third International Conference on Intelligent Sensors, Sensor Networks and Information Processing* (ISSNIP), 2007.
27. Ma Di and Er Meng Joo, "A survey of machine learning in wireless sensor networks from networking and application perspectives," in *Information, Communications & Signal Processing*, 2007 6th international conference, 2007, pp. 1–5.
28. V. Kulkarni, A. Forster, and G. Venayagamoorthy, "Computational Intelligence in wireless sensor networks: A survey," *Communications Surveys & Tutorials*, vol. PP, 2010, pp. 1–29.
29. A. Forster, "Teaching networks how to learn," PhD diss., Faculty of Informatics, University of Lugano, Switzerland, May 2009.
30. G. Vijay, "Cognitive wireless sensor networks," PhD diss., Queen's University, Kingston, Canada, 2012 (forthcoming).

2

Cognitive Radio Networks and Dynamic Spectrum Access

2.1 Introduction

Two fundamental trends have jointly led to cognitive radio (CR) technology. The first one is the gradually increasing utilization of adaptation techniques in the transmitter/receiver design—for example, channel estimation, power control, adaptive modulation, and many other forms of adapting the radio unit parameters in response to stimulus from the surrounding environment. This trend coincided with the emergence of the concept of software-defined radio (SDR) as a communication unit that has most of its functionalities performed in software. In other words, SDR is a radio unit that is capable of easily supporting different technologies, standards, and environmental conditions. The second trend is the growing belief that we are actually approaching the physical limit of the radio spectrum, especially under the 3 GHz bands. This has led some spectrum regulation agencies, such as the Federal Communications Commission (FCC), to start revising their spectrum allocation strategies. Fortunately, this unveiled the fact that the spectrum is highly underutilized in some bands below the 3 GHz range. This discovery motivated the search for novel means of spectrum access such that spectrum utilization efficiency can be enhanced. These somehow isolated development trends met in a single point where their capabilities and promises could simultaneously be achieved. That point—CR technology—is a promising field that can take the wireless communications to new dimensions of flexibility, efficiency, and seamless communication.

2.1.1 History of Cognitive Radio

The concept of CR was first described by J. Mitola in his PhD dissertation (Mitola 2000) as means to enhance the flexibility of wireless communications by increasing the awareness of the radio unit about its surrounding environment and about itself. Shortly after that, the FCC saw in this technology a solution to the spectrum underutilization problem. The Spectrum Policy

Task Force (SPTF) was formed to review the spectrum allocation policies and recommended CR as a remedy to this problem in 2003 (*Notice of Proposed Rule Making* 2003). Among the first contributions in the CR literature were the work of S. Haykin (2005) and the publication series of the Berkeley Wireless Research Center (BWRC) (Cabric, Mishra, and Brodersen 2004; Cabric et al. 2005). Haykin (2005) laid the ground for a signal processing and communications realization of CR technology. Similarly, Cabric, Mishra, and Brodersen (2004) and Cabric et al. (2005) addressed some fundamental issues with the physical realization of the technology. In these early contributions, two main tasks for CR were identified, namely:

- Spectrum sensing to detect vacant spectrum bands (also known as spectrum holes)
- Spectrum access to efficiently utilize these spectrum holes

Spectrum sensing in CR has some similarity to carrier sensing (CS) in random access wireless networks. Both need to sense the transmission medium before initiating the communication. However, there are two main differences between them. First, in CS, there are few access channels. Hence, if they are all busy, transmission will be postponed. In CR, on the other hand, the number of access channels is large, and thus access opportunities are large. Secondly, in CS, all nodes know the exact characteristics of the transmission carrier; hence the sensing process is not that difficult. However, CRs are assumed to detect unknown or partially known communication signals to identify spectrum holes, which is a major challenge.

Spectrum access attracted less attention in the literature than did spectrum sensing. This is because in normal circumstances, once a CR detects a spectrum hole, it can access it like a conventional communication transmitter. However, in the past few years, considerable attention has been paid to improve the spectrum utilization by benefitting from multiple-input/multiple-output (MIMO) and cooperative diversity techniques, and using more than one spectrum hole at a time.

2.1.2 MIMO and Cooperative Diversity Techniques

MIMO techniques refer to a family of techniques wherein multiple antennas are used to enhance system capacity and to mitigate channel impairments. These include beamforming, spatial multiplexing, and many others. Beamforming is a signal processing technique used to direct transmission and/or reception in MIMO systems. It can either be designed to maximize the communication throughput by enhancing the signal quality or to reduce interference. In general, beamforming can be used to achieve spatial selectivity by using either adaptive or fixed beam patterns to direct the transmission or the reception beam. On the other hand, spatial multiplexing is used to simultaneously transmit messages from different users by using different transmit antennas. The main advantage of spatial multiplexing is to increase the data rate achievable over the same time and frequency resources (Larsson and Stoica 2003).

Motivated by the successes of MIMO techniques, cooperative diversity techniques started recently to gain considerable attention in various wireless applications. In particular, cooperative diversity aims at overcoming some of the limitations of wireless systems, such as transmission range and communication reliability. The basic idea behind cooperative diversity is that when a source transmits a message to a destination, this message is also received by other terminals in the network that are often referred to as relays or partners. These relays process the message and retransmit it to the destination. The destination then combines the messages coming from the source and the partners. In doing so, the system provides spatial diversity by taking advantage of the multiple receptions of the same message. This idea is illustrated in Figure 2.1. Another advantage of cooperative diversity is the dramatic suppression of interference between communications terminals through its distributed spatial processing (Sendonaris, Erkip, and Aazhang 2003a, 2003b).

Recently, interest has increased in benefiting from the advantages of MIMO systems in CR networks. As we shall see in subsequent sections, MIMO systems can be used in both parts of the dynamic spectrum access (DSA) process: the sensing process and the spectrum access process. This interest has opened new research directions in the past few years. New challenges arise, such as the design of efficient cooperative sensing methods, while some conventional challenges such as the design of efficient beamforming techniques need to be reconsidered.

The remainder of this chapter is organized as follows. In Section 2.2 the concept of spectrum awareness is introduced and analyzed. The main challenges and methods for spectrum sensing are addressed therein. Section 2.3

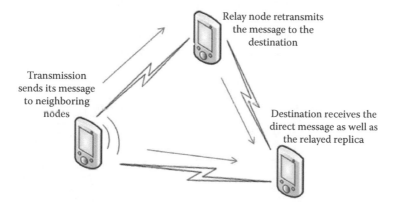

FIGURE 2.1
Illustration of cooperative diversity.

is devoted to cooperative spectrum sensing, which is a powerful tool for enhancing the sensing performance in CR networks, while Section 2.4 addresses dynamic spectrum access with an emphasis on the cooperative communications aspect of it. Finally, conclusions are drawn in Section 2.5.

2.2 Spectrum Awareness

Obtaining accurate information about the status of the available frequency bands is a task lying at the core of CR technology. In a broad sense, this task is referred to as spectrum awareness. The methods used to obtain spectrum awareness can be classified as shown in Figure 2.2. For similar classifications and comprehensive surveys, see Zhao and Sadler (2007), Yucek and Arslan (2009), Arslan (2007, Chapter 9), and Fitzek and Katz (2007, Chapter 18).

In Figure 2.2, spectrum awareness is divided into two main categories: passive awareness and active awareness. In the first category, the CR relies on other entities or agents, or even on direct communication with the primary user (PU) itself to obtain knowledge about the status of the spectrum bands it plans to utilize. This category encompasses a wide range of techniques, including the beacons approach (Hulbert 2005), the secondary markets (*Principles for Promoting* 2000), the policy-based information exchange (Mangold et al. 2004), and many others (such as Cave, Doyle, and Webb 2007). For instance, an independent sensing network that keeps monitoring the spectrum instead of the CR network was proposed in Han, Fan, and Jiang (2009). When a CR attempts to transmit data, it negotiates with the closest sensing node to know the available bands. Once a CR node is assigned a spectrum hole, it can start the transmission. Consequently, CRs do not need to monitor the spectrum themselves; rather, they seek access from the sensing network, which is continuously monitoring the spectrum.

Almost all passive awareness techniques presume that PUs are willing to share some information about their communications with the CR network.

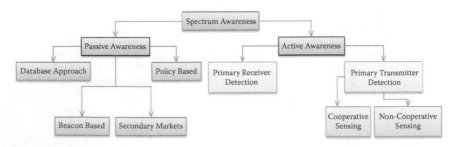

FIGURE 2.2
Classification of spectrum awareness techniques.

This information helps CRs to know the status of the spectrum. This information could be a periodically broadcasted pilot signal that makes the sensing process easier. PUs can broadcast the transmission characteristics or even the spectrum utilization patterns. This presumption is one of the reasons halting the deployment of CRs based on passive spectrum awareness. It is also a growing cause of discomfort for the primary network operators.

The active spectrum awareness category, on the other hand, makes no presumptions on the PUs' network. Active spectrum awareness is referred to in the literature as spectrum sensing because the spectrum sensing techniques rely solely on their sensing capabilities to obtain the spectrum status. Furthermore, unlike passive awareness techniques, spectrum sensing techniques assume no modifications to the PUs' transmissions. In other words, it is a friendly deployment that relies solely on the CR nodes' potentials to attain coexistence between the two networks.

Spectrum sensing is subdivided into two types depending on the targeted objective, as shown in Figure 2.2. These are based on detecting primary receivers and detecting primary transmitters. Of the two types, detecting primary transmitters attracted most of the attention in the literature since transmitters are easier to detect than receivers due to the transmission power radiation.

An example of receiver detection is the detection of TV sets. TV sets are receivers that have no transmission capabilities. This detection problem was encountered in detecting spectrum holes within the TV bands to be used in the IEEE 802.22 standard as shown in Figure 2.3.

For instance, Wild and Ramchandran (2005) utilized the local oscillator leakage power of the TV sets to detect the presence of TV sets. This approach, though, forces spectrum reusability to the limit, is impractical, and is difficult to implement. The reasons for this are as follows. First, detecting primary receivers is either impossible when they have no radiations at all or is limited to very short distances due to the low radiated power (Wild and Ramchandran 2005). Second, even if the CR is assumed to know the

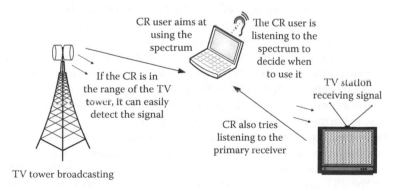

FIGURE 2.3
Detection of TV signals.

absence of all primary receivers, it still cannot use the bands as long as the primary transmitter (the TV tower in the IEEE 802.22 case) is broadcasting. This is obvious since the CR signal will coexist with the primary signal. Consequently, both signals will jam each other, and the receiver with the lower signal power will fail to detect its signal.

On the other hand, detecting primary transmitters has been thoroughly studied in the literature. This type of detection can be performed either locally, where each CR user senses the spectrum to make its own decision without collaborating with others, or globally, where a group of CR users collaborate in sensing a particular frequency band. The former method is referred to as noncooperative sensing, while the latter method is referred to as cooperative sensing. Obviously, cooperative sensing outperforms noncooperative sensing because it benefits from collaborative efforts of all individual CRs to make better decisions about the status of the spectrum bands (Mishra, Sahai, and Broderson 2006).

A brief look at the main challenges facing the development of efficient spectrum sensing methods is next, followed by a study of the different transmitter sensing methods proposed in the literature.

2.2.1 Spectrum Sensing Challenges

CRs are envisioned to be able to operate with minor knowledge about the existing PUs. This vision covers a scale of varying prior information about the PUs. At one end of the scale, CRs are assumed to have no knowledge about the PU's communication. This case is the most challenging when the design of efficient spectrum sensing methods is considered. At the other end, CRs are assumed to know the exact signals available and even be synchronized with them. Obviously, this is a less challenging situation that facilitates the design process. Between these two ends, the difficulty of the design process varies with the amount of information available.

In general, there are some fundamental challenges facing the development of efficient spectrum sensing methods. Some of these challenges are technology dependent, such as the need for flexible radio frequency (RF) front ends and high-resolution analog-to-digital converters (ADC), while others depend on the nature of the PU's signal, the status of the wireless channel, and many other factors. These types of challenges are considered here.

The first challenge is the hidden PU problem (also referred to as hidden terminal problem). This is a classical problem in random access wireless networks. It happens in CR technology when a CR fails to detect an active primary transmitter communicating with a primary receiver located in the transmission range of the CR. This situation is illustrated in Figure 2.4.

When this situation happens, the CR decides to transmit in the same band used by the PU, and hence causes intolerable interference to the primary receiver. This problem may be caused by a multiplicity of factors, such as deep fading and shadowing experienced by the CR while sensing the band.

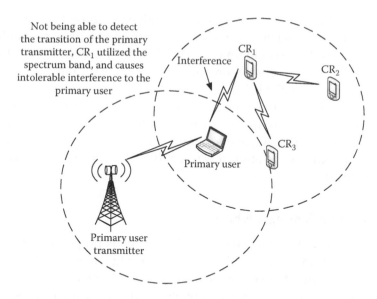

Not being able to detect the transition of the primary transmitter, CR_1 utilized the spectrum band, and causes intolerable interference to the primary user

Interference

CR_1

CR_2

CR_3

Primary user

Primary user transmitter

FIGURE 2.4
Illustration of the hidden PU problem in CR systems.

It can also be caused by the presence of some physical barriers, like buildings or mountains. This problem may be solved using cooperative sensing, as we shall shortly see in Section 2.3.

The second challenge is encountered when the PU is using spread spectrum signals. These signals were originally designed to be immune to jamming, thus they are inherently difficult to detect. For CR to detect these signals, it needs to rely on some of the passive spectrum awareness techniques mentioned above. Alternatively, the CR can utilize a hybrid passive–active spectrum sensing technique such that the PU shares the possible spreading codes (in direct sequence-based spread spectrum) with the CR or the frequency hopping sequence (in frequency-hopping-based spread spectrum) with the CR. Fortunately, spread spectrum signals are mainly used for cellular communications. FCC measurements (*Spectrum Policy Task Force Report* 2002) showed that the cellular communications bands are among the most heavily occupied bands, and they are difficult to consider for opportunistic access.

The third challenge is the sensing time and the evacuation mechanism. The CR needs to search for an idle frequency band and use it as long as the PU is not using it. Once the PU reclaims its band again and starts transmission, the CR needs to evacuate the band immediately. If the CR is using spectrum sensing, then it needs to immediately sense the presence of the PU and evacuate the band.

Despite being crucially important, this problem has been lightly considered in the literature. The main contributions to solve this problem were based on medium access control (MAC) layer protocols (see Xiao and Hu

2009, Chapter 8). Some of the suggested solutions propose having multiple opportunistic bands where the CR can move from one band to another in order to ensure reliable communication for the CR. However, these solutions aimed at guaranteeing reliable communication for the CR rather than reducing the interference to the PU.

There are many other challenges encountered in particular scenarios—for example, the effect of fading on the reporting channels in cooperative sensing, the effect of fading and shadowing on the performance of sensing methods, the complexity/robustness tradeoff, and many others (Arslan 2007; Xiao and Hu 2009).

2.2.2 Spectrum Sensing Methods

There is a multiplicity of spectrum sensing methods proposed in the literature, each assuming some level of prior knowledge about the PU's signal. The most commonly studied methods are energy detection, cyclostationarity detection, covariance detection, and others, such as matched filters. In this section, we focus on the first three methods and briefly describe the others.

2.2.2.1 Energy Detection

This is an optimal detection method when no prior information about the primary signal is available. Hence, it is the most widely used sensing method in the literature (Cabric, Mishra, and Brodersen 2004; Ben Letaief and Zhang 2009; Ganesan and Li 2007a, 2007b). Energy detection, as the name suggests, decides on the presence of the PU based on an estimate of the energy present in some spectrum band. Mathematically speaking, a decision metric is calculated as:

$$y = \sum_{k=1}^{2TW} |r_k|^2,$$

(2.1)

where $\{r_k\}_{k=1}^{2TW}$ are samples of the received signal over the frequency band of W Hz observed over T seconds. This decision metric is then compared to a threshold λ to decide about the presence or absence of the PU,

$$y \underset{\mathcal{H}_0}{\overset{\mathcal{H}_1}{\gtrless}} \lambda,$$

(2.2)

where the two hypotheses $\mathcal{H}_0$ and $\mathcal{H}_1$ are defined as primary signal absent and primary signal present, respectively. The received signal—assuming *additive white Gaussian noise* (AWGN) channel only—is

$$r(t) = \begin{cases} s(t) + n(t), & \mathcal{H}_1, \\ n(t), & \mathcal{H}_0, \end{cases} \tag{2.3}$$

Choosing the decision threshold is a major challenge in the design of the energy detector. This threshold depends on the variance of the additive noise $n(t)$. Hence, a good estimate of the noise variance is needed to properly design this detection method. In fact, it was shown in Sahai, Hoven, and Tandra (2004) that the performance of the energy detection severely degrades when the estimate of the noise variance is not accurate. Furthermore, it was also shown that a total of $O(1/SNR^2)$ samples are needed to guarantee a desired performance level. In other words, the observation time T needs to be prolonged as the signal to noise ratio (SNR) of the PU decreases.

2.2.2.2 Cyclostationarity

In this method, the cyclostationarity features of the primary signal are exploited to detect the presence of PUs. These features are due to the periodicity of either the signal or its statistics, such as the mean and the autocorrelation function. When the primary signal exhibits strong cyclostationarity features, the CR can detect it even under low SNR conditions (Yucek and Arslan 2009). Mathematically, the cyclostationarity detector does the following (Ben Letaief and Zhang 2009):

1. Calculates the cyclic autocorrelation function (ACF) of the received signal, $x(t)$, as: $R_{xx}^{\alpha}(\tau) = E\left[x(t+\tau)x^*(t-\tau)e^{-j2\pi\alpha t} \right]$, where α is the cyclic feature.
2. Calculates the cyclic spectral density (CSD) as the discrete Fourier transform of the ACF: $S(f,\alpha) = \sum_{\tau=-\infty}^{\infty} R_{xx}^{\alpha}(\tau)e^{-j2\pi f\tau}$.
3. Searches for the unique cyclic frequency corresponding to the peaks of the CSD (a two-dimensional function).

This completes the detection process.

The main advantage of this detection method over the energy detection is its immunity to the noise signal. In other words, this method is robust to the uncertainty of the noise power since the noise signal is a wide-sense stationary process and thus has no periodicity in its autocorrelation function. However, this method requires additional information about the primary signal. In particular, it requires prior knowledge about the modulation format to identify the cyclic frequency. In addition, this method requires longer observation time and additional implementation complexity (Yucek and Arslan 2009; Ben Lataief and Zhang 2009). This sensing method was tested in the development of the IEEE 802.22 standard in Han et al. (2006). The results showed superior performance gains over the energy detection at low SNR levels. Furthermore, in Lundén et al. (2009), this method was also extended into two dimensions. The first is by considering multiple cyclic frequencies to distinguish between primary signals and other CRs. In the second, a collaborative scheme to overcome the hidden terminal problem and the channel impairments effects was proposed. In addition, by utilizing the uniqueness of the cyclic frequencies for different signals, Oner and Jondral (2007) proposed a strategy to extract the channel allocation information for the CRs. Oner and Jondral (2007) considered the case where the PU has a global system for mobile communications (GSM) signal while the secondary is a wireless local area network (WLAN) system using orthogonal frequency division multiple access (OFDMA).

2.2.2.3 Covariance Detection

Benefiting from the difference in the structure of the covariance matrix of the received signal when the primary signal is present or absent, the covariance-based spectrum sensing method was proposed in Zeng and Liang (2007). In this method, estimates of the off-diagonal elements of the covariance matrix are compared with some detection thresholds. Mathematically, if the received signal $r(t)$ is written as in Equation (2.3) above, then by observing the signal for L consecutive symbol durations, we can define the vectors $\mathbf{r}[k] = [r[k], r[k-1], \ldots, r[k-L+1]]^T$, $\mathbf{s}[k] = [s[k], s[k-1], \ldots, s[k-L+1]]^T$, and $\mathbf{n}[k] = [n[k], n[k-1], \ldots, n[k-L+1]]^T$. Hence, we can write:

$$\mathbf{r}[k] = \begin{cases} \mathbf{s}[k] + \mathbf{n}[k], & \mathcal{H}_1, \\ \mathbf{n}[k], & \mathcal{H}_0, \end{cases} \tag{2.4}$$

If we further define the correlation matrix of the received signal $\mathbf{r}[k]$ as $\mathbf{R}_r = E\left[\mathbf{r}[k]\mathbf{r}^H[k] \right]$, where $(\cdot)^H$ is the Hermitian operator, then we can write:

$$\mathbf{R}_r[k] = \begin{cases} \mathbf{R}_s[k] + \mathbf{R}_n[k], & \mathcal{H}_1, \\ \mathbf{R}_n[k], & \mathcal{H}_0, \end{cases} \tag{2.5}$$

where $\mathbf{R}_s[k]$ and $\mathbf{R}_n[k]$ are defined similar to $\mathbf{R}_r[k]$ above. Knowing that the correlation matrix of the additive noise, $\mathbf{R}_n[k]$, is a diagonal matrix, the covariance detection method simply compares the off-diagonal elements of $\mathbf{R}_r[n]$ with some threshold value. If the signal is present, the off-diagonal elements should have nonzero values exceeding the threshold value; otherwise, it should be zeros. This idea was further extended in Zeng and Liang (2009) using the eigenvalues decomposition of the covariance matrix and the change happening to its eigenvalues when the primary signal is present or absent.

The covariance detection methods have the advantage of being robust to noise variance similar to the cyclostationarity detection method. However, they also rely on accurate estimations of the covariance matrix, which requires prolonged observation time. In addition, setting the thresholds for comparison is a critical step.

2.2.2.4 Other Detection Methods

There are other spectrum sensing methods proposed in the literature. These include the wavelet-based detection method (Tian and Giannakis 2006). This method was also used in Hur et al. (2006) to build a two-stage sensing method. Another famous method is the matched filter detection. This is the optimal coherent detection method when perfect knowledge about the primary signal is available. However, implementing this method requires time synchronization with the primary signal. Achieving this is too difficult when there is no direct interaction (communications or collaboration) between the CR and the PU. The interested reader can refer to Arslan (2007) and Fitzek and Katz (2007), where comprehensive discussions about this detection method are given.

All detection methods discussed so far observe some spectrum band for a period of time before making a decision about the availability of a PU in this band or not. An alternative approach was recently proposed in Lai, Fan, and Poor (2008), who restate the sensing problem as a change detection problem wherein the CR keeps listening to the spectrum band searching for instances of change in the behavior of the PU. Based on this formulation, a quick detection method that minimizes the detection time was designed. This approach of change detection is suitable if the CR is targeting a specific set of bands to opportunistically utilize some of them once the PUs evacuate them. However, it lacks the needed flexibility to search for spectrum holes in the wideband portion of the spectrum.

Using the same mathematical tools, Li, Li, and Dai (2008) considered the special case of detecting PUs' activities, particularly TV channels. The problem was customized to allow CRs to detect a sinusoidal pilot signal sent by PUs upon resuming their transmission.

In conclusion, many spectrum sensing methods have been studied in the literature. Some of them are included in survey papers, such as Zhao and Sadler (2007) and Yucek and Arslan (2009), while others were mentioned in the context of cooperative sensing, such as Ben Letaief and Zhang (2009) and Quan et al. (2008).

2.2.2.5 Comparison between Different Sensing Methods

We have thus far briefly addressed some of the basic spectrum sensing methods. It was manifest that different detectors are applicable in different scenarios. Of all the detectors, the energy detector is the simplest to implement.

It is capable of detecting arbitrary primary signals without requiring any prior information. The performance of energy detectors over various fading channels with and without diversity combining was studied in Kostylev (2002); Digham, Alouini, and Simon (2003, 2007); Herath and Rajatheva (2008); and Herath, Rajatheva, and Tellambura (2009).

To improve the detection performance, it is important for the detection method to be able to differentiate the primary signal from interference and noise. In practice, however, this differentiation can only be realized if some a priori knowledge of the primary signal is known to the CR. Depending on what information the CR has about the primary signal, different detectors can be applied in different scenarios. For instance, a covariance-based detector is suitable when the primary signal is known as correlated. On the other hand, a cyclostationarity detector is suitable when the period of the primary signal is known. Furthermore, a matched filter detector is suitable when the pilot signal of the primary system is known, such as in the case of TV bands. In general, the more the CR user knows about the primary signal, the better the detector works (see Figure 2.5).

Since CR technology is inherently a wireless communications technology, the effect of channel impairments on the performance of the sensing method cannot be ignored. The impairments of wireless channels include small-scale fading, shadowing, and path loss. These factors degrade the performance of the sensing method. The performance degradation on most sensing methods has been studied in the literature either explicitly, as the case for the energy detectors (Kostylev 2002; Digham, Alouini, and Simon 2003, 2007), or implicitly as in the eigenvalues-based methods (actually by approximation as well), as in Zeng and Liang (2009) and the cyclostationarity method (Lundén et al.

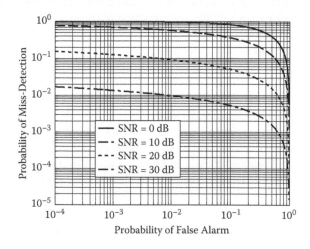

FIGURE 2.5
Performance of energy detector over Rayleigh fading channels with different SNR levels, SNR = 0, 10, 20, and 30 dB.

2009). In fact, since the autocorrelation function of the received signal is estimated to get the cyclic power spectrum, the performance analysis processes become too difficult.

2.3 Cooperative Sensing

The hidden terminal problem can be addressed by allowing multiple CRs to cooperate in spectrum sensing (Cabric, Mishra, and Brodersen 2004). This cooperative sensing is analogous to distributed detection in wireless sensor networks (WSNs). In WSNs, each node makes its local decision and forwards it to a common fusion center (FC) or central node that makes the final decision based on some fusion rule (Ben Letaief and Zhang 2009). A major difference between these two applications is in the coverage area of the network. While WSNs cover smaller areas (most of the time), CRs are presumed to cover wider areas, and thus designing cooperative sensing is more challenging. On the other hand, CRs are not battery powered like the WSN nodes. Thus, they can benefit from sophisticated and advanced signal processing techniques in designing the cooperative sensing.

Most of the contributions in the literature focus on the narrowband cooperative sensing wherein the CRs collaborate to sense a particular narrowband spectrum. Wideband sensing, on the other hand, allows the CRs to monitor a group of narrowband spectrum portions so that the CR network has a lot of options when intending to access the spectrum. In the following sections, these two variants of cooperative spectrum sensing will be studied.

2.3.1 Narrowband Cooperative Sensing

Cooperative sensing can either be centralized or distributed (Ma, Li, and Juang 2009) depending on the decision-making process. In centralized cooperative sensing, all local decisions are sent to a FC that makes the final decision and informs all the CRs thereof, as shown in Figure 2.6. In other words, centralized cooperative sensing works as follows (Ben Letaief and Zhang 2009):

1. Each CR makes its local sensing using some sensing method as discussed above. In almost all papers addressing cooperative sensing, the energy detection is assumed (Ganesan and Li 2007a, 2007b).
2. All users forward their local decisions to the FC.
3. The FC fuses all the data according to some fusion rule to make the final decision. The final decision is then sent to all CRs.

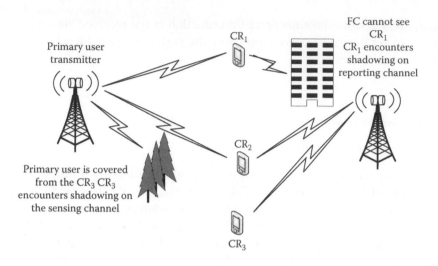

FIGURE 2.6
Cooperative sensing.

In distributed cooperative sensing, on the other hand, neighboring CRs share their sensing information; each CR makes its final decision using its own fusion rule.

Let us assume that a group of K CRs senses a particular spectrum band and forwards their decisions to a FC. Let the decision of each node be denoted by d_i, $i = 1, \ldots, K$. When the PU is present, $d_i = 1$, and when it is absent (that is, a spectrum hole is available), $d_i = 0$. Furthermore, let R denote the final decision made by the FC. Thus we can write

$$R = \sum_{i=1}^{K} d_i \begin{cases} \geq n & \mathcal{H}_1 \\ < n & \mathcal{H}_0 \end{cases} \tag{2.6}$$

where the FC chooses between two hypotheses, $\mathcal{H}_1$ where the PU is present and $\mathcal{H}_0$ when it is absent and a spectrum hole is available. If at least n CRs decided that a PU is present, the FC chooses $\mathcal{H}_1$; otherwise it chooses $\mathcal{H}_0$. The selection of the voting threshold n defines the decision rule used. There are three commonly used decision rules (Ben Letaief and Zhang 2009): the AND rule, ($n = K$); the OR rule, ($n = 1$); and the majority rule, ($n > K/2$). The AND rule chooses $\mathcal{H}_1$ if all the CRs choose it. Apparently, this rule is the most aggressive rule in detecting spectrum holes. On the other hand, the OR rule chooses $\mathcal{H}_1$ if at least one CR chooses it. Thus, this rule is the most conservative rule. Finally, the majority rule strikes a balance between these rules by choosing $\mathcal{H}_1$ if the majority of the users choose it. A performance comparison between these three rules is shown in Figure 2.7. It is manifest that the detection probability of the OR rule is superior to the other rules

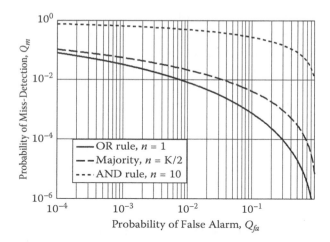

FIGURE 2.7
Centralized cooperative spectrum sensing performance with different fusion rules over fading channels with SNR = 10 dB and $K = 10$.

i.e., achieving the lowest miss-detection for the same SNR and false alarm probability constraint.

The model in Equation (2.6) is indeed an oversimplified one. It does not show the effects of fading and shadowing on the performance of cooperative sensing. In fact, the performance of cooperative sensing depends on three factors: the status of the sensing channels between the PU and the CRs, the status of the reporting channels between the CRs and the FC, and the decision rule. The effect of fading in the reporting channels on the cooperative sensing was studied in Ghasemi and Sousi (2005). Special emphasis was given to the case of spatially correlated shadowing. On the other hand, the effect of fading in the sensing channels on some of the sensing methods has been studied in the literature, as mentioned in Section 2.2.2. The effect of these two factors on the performance of cooperative sensing can be mitigated using spatial diversity. There is more than one way to achieve spatial diversity, as mentioned in Section 2.1.2. These include using MIMO systems and using cooperative diversity. Using MIMO systems was studied in Digham, Alouini, and Simon (2003) to enhance the performance of the energy detection over various fading channels, while the use of cooperative diversity was recently studied in Ben Letaief and Zhang (2009). The authors used a distributed Alamouti space time code to make couples of CRs relay their local decisions to the FC. Using this cooperation scheme, the error rate over the reporting channels was significantly reduced. We study this topic further in Section 2.3.2.

The third factor affecting the performance of cooperative sensing is the decision rule used. In the model above, the FC follows the suboptimal voting scheme to make the final decision. This scheme achieves decent performance at a reduced overhead in the reporting channels. An alternative decision-making rule was proposed in Quan et al. (2008), who extended the optimal (when no prior information about the PUs is available) energy detection method to the FC. In other words, each CR forwards its measurements (also referred to as summary statistics) to the FC, which optimally combines them and makes the final decision. Mathematically, each CR forwards its summary statistics $T_i(\mathbf{x}_i)$, $i = 1, \ldots, K$, to the FC, which calculates the likelihood ratio $L(\mathbf{y})$, $\mathbf{y} = [T(\mathbf{x}_1), \ldots, T(\mathbf{x}_K)]^T$, as

$$L(y) = \frac{P(\mathbf{y}\,|\,\mathcal{H}_1)}{P(\mathbf{y}\,|\,\mathcal{H}_0)} \underset{\mathcal{H}_0}{\overset{\mathcal{H}_1}{\gtrless}} \gamma \tag{2.7}$$

and the result is then compared to a threshold value γ to make the final decision. Obviously, the same difficulties faced by the energy detectors, especially choosing the decision threshold, will also be faced in this scheme. However, due to the complexity in designing a detection method based on the likelihood ratio, Quan et al. propose a reduced complexity decision statistic based on a weighted linear combination of the summary statistics. This simplified decision statistic uses the following rule

$$L_l(\mathbf{y}) = \mathbf{w}^T \mathbf{y} \underset{\mathcal{H}_0}{\overset{\mathcal{H}_1}{\gtrless}} \gamma \tag{2.8}$$

to make the final decisions. The weights vector $\mathbf{w}$ can be chosen to account for the contribution of the individual CRs. For instance, it can be a function of the SNR; CRs with large SNR are weighted more than CRs with small SNR. It was shown in Quan et al. (2008) that the performance of this suboptimal decision rule is indeed comparable to that based on the likelihood ratio if the weights vector was properly chosen.

2.3.2 Wideband Cooperative Sensing

The ability of the CR to sense a wideband spectrum is an important topic studied in the spectrum sensing literature (Tang 2005). The initial proposals for wideband sensing assumed a flexible radio frequency (RF) front end that allows the CR to sweep over a wide range of the spectrum (Cabric, Mishra, and Brodersen 2004), as shown in Figure 2.8. However, this setup is slow and inflexible (Quan et al. 2008) and is limited by the availability of a flexible RF front end in hardware. Another issue with this proposal is the sensing order—that is, where shall the sensing method start, what band shall be sensed next, and how to optimally search for a spectrum hole in a wide range of spectrum.

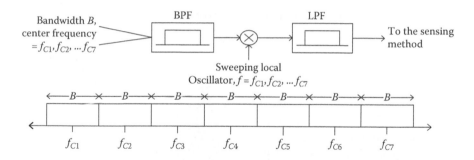

FIGURE 2.8
Wideband spectrum sensing using sweep circuit.

Some of these questions were actually addressed in the literature. For instance, in Luo et al. (2009), the sensing process was divided into two stages: a coarse sensing stage and a fine sensing stage. In the first step, a wide spectrum range is tested using a low complexity sensing method such as energy detection. If a spectrum hole is detected, the second stage is invoked. In this stage, a high-accuracy sensing method such as cyclostationarity or matched filter is used to search for the exact location of the spectrum hole. Analytical results showed enhanced detection agility when the number of spectrum holes is small compared to the conventional one-stage sensing.

An alternative solution to wideband spectrum sensing was proposed in Quan et al. (2008) using cooperative sensing. It is assumed that each CR observes a bank of K narrowband spectrum bands and forwards its summary statistics about all the bands to the FC. The FC makes a linear combination of these statistics and compares it to a decision threshold [see Equation (2.8)]. Let us denote the summary statistics at the FC for the kth narrowband by $\mathbf{Y}_k = [T_{k,1}, T_{k,2}, \ldots, T_{k,M}]^T$; the FC uses the following decision rule

$$z_k = \mathbf{w}_k^T \mathbf{Y}_R \underset{\mathcal{H}_{0,k}}{\overset{\mathcal{H}_{1,k}}{\gtrless}} \gamma \tag{2.9}$$

Similar to the linear combination solution in Equation (2.8), the choice of the weighting vector $\mathbf{w}_R$ needs some design criteria. While in the narrowband case, Quan et al. (2008) used the SNR criteria, here they designed the weight vector such that the throughput of the CR network is maximized while meeting an interference limit on the adjacent PUs' network.

2.4 Dynamic Spectrum Access

Having understood the spectrum sensing process with its various aspects, let us now look at the approaches for spectrum access. There are two spectrum

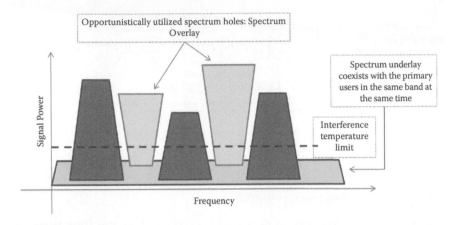

FIGURE 2.9
Dynamic spectrum access.

access approaches studied in the literature: spectrum overlay and spectrum underlay (Zhao and Sadler 2007). These two schemes are shown in Figure 2.9.

In spectrum overlay, CRs sense some frequency bands; once a spectrum hole is detected, the CR uses it for transmission. This approach protects the rights of PUs by using the band only when it is not used by the PU and evacuating it once the PU resumes transmission. From the PU's perspective, there are two critical concerns regarding this approach:

- The CR's ability to reduce the miss-detection as much as possible. Missed detections are occasions wherein a band is deemed empty while it is actually busy.
- The CR's ability to evacuate the band as soon as the PU resumes transmission.

The first concern urges developing high-accuracy sensing methods and efficient cooperative sensing schemes as discussed previously. Although the second concern might be considered an extension of the first one, designing sensing methods for the evacuation is different from designing the regular sensing methods. While the latter aims at achieving reliable sensing, the former aims at reducing the detection time as much as possible. This difference was recently observed in Lai, Fan, and Poor (2008), as mentioned in Section 2.2.2.4, wherein a detection method that minimizes the detection time by continuously observing the used band was designed using the quickest detection framework.

On the other hand, spectrum underlay allows CRs to access the spectrum whenever they need to as long as they are not causing harmful interference to the neighboring PUs. To enable this approach, a reliable interference control mechanism needs to be employed by all the CRs. This mechanism is composed of the following steps:

1. Detect all the PUs in the neighborhood and estimate the distance to each one.
2. Decide the interference limit for each PU.
3. Calculate the transmission power such that the interference limits are not exceeded at any PU.

The FCC proposed the interference temperature (IT) as a measure of the interference at the primary receivers (*Establishment of Interference Temperature Metric* 2003). Despite the fact that this approach does not need reliable spectrum sensing and that CRs can access the spectrum whenever needed, the development of this approach poses some serious challenges:

• The CR needs to continuously monitor all the PUs in its transmission range. This is a challenging process that needs advanced signal processing techniques such as estimating the angle of arrival and estimating the power received from each PU. Furthermore, this process needs to be continuously performed when the PUs are mobile.

• The IT needs to be specified for each neighboring PU. In fact, the concept of IT attracted little attention in the literature since estimating it is practically difficult (Haykin 2005). Furthermore, it was shown in Clancy (2007) that there is more than one unique way to define the IT, and that each one affects the achievable throughput for the CR as well as the performance of the neighboring PUs differently. The appropriate power control method used by the CRs needs to be determined. In the underlay approach, CRs adjust their transmission power to meet the IT limit at all neighboring PUs. However, this can affect the performance of the CR network significantly.

For these challenges, the literature focused more on the spectrum overlay approach for spectrum access. In fact, even the FCC themselves withdrew their proposition regarding the IT (*Re: Establishment of an Interference Temperature Metric* 2007). However, this idea is still being studied by some researchers for possible spectrum access (Zhang and Liang 2008; Zhang, Liang, and Xin 2008).

2.4.1 MIMO Systems for Spectrum Access

In the underlay approach, CRs are allowed to use the same spectrum band as a PU at the same time and geographic area provided that this does not cause harmful interference to the PU. Consequently, designing the transmission

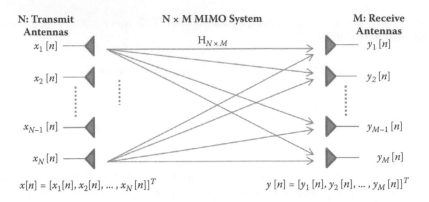

FIGURE 2.10
MIMO system.

matrix for a CR MIMO system becomes a challenging problem. The CR needs to maximize its throughput while maintaining acceptable interference levels to the neighboring PUs.

Consider the $N \times N$ MIMO system shown in Figure 2.10. The transmission vector at the nth time slot is $\mathbf{x}[n] = \left[x_1[n], x_2[n], \ldots, x[N]\right]^T \in \mathbb{C}^{N \times 1}$. Selecting the symbols in this vector decides the type of MIMO system used. If the symbols are identical (that is, $x_1[n] = x_2[n] = \cdots = x_{N-1}[n]$), then the system is achieving space diversity at the transmitter; if the symbols belong to different users, the system is a spatial multiplexer. Furthermore, if the symbols are identical and are weighted differently by some weighting vector to cope with the channel information, the system is a transmit beamforming. Finally, if the received symbols $\mathbf{y}[n] = \left[y_1[n], y_2[n], \ldots, y_M[n]\right]^T$ are also weighted by some weighting vector $\mathbf{w}$, then the system is a transmit/receive beamforming.

In general, the design process of these beamforming vectors is an optimization process that can be written as (Boyd and Vandenberghe 2004):

$$\begin{aligned} \text{minimize} \quad & f_0(\mathbf{v}, \mathbf{w}) \\ \text{subject to} \quad & f_i(\mathbf{v}, \mathbf{w}) \le b_i, \quad i = 1, \ldots, m. \end{aligned} \tag{2.10}$$

The objective or the cost function $f(\mathbf{v}, \mathbf{w})$ decides the design criterion. It can be the error rate, the average SNR, or another cost function, as discussed in Palomar, Cioffi, and Lagunas (2003). Similarly, the constraint functions $\left\{f_i(\mathbf{v}, \mathbf{w})\right\}_{i=1}^m$ can take different forms. The most commonly used ones are the transmit power constraint and the interference constraint. Under the umbrella of CR, a similar optimization shall be solved to design the beamforming vectors. However, the cost and constraint functions depend on the scenario considered. For instance, in Zhang and Liang (2008), a pair of CRs was assumed to access a particular spectrum occupied by a group of PUs, as shown in Figure 2.11.

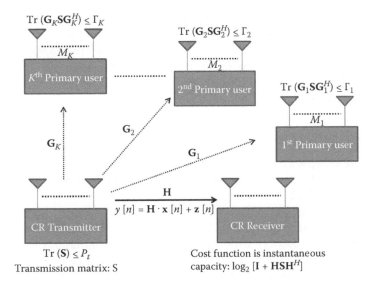

FIGURE 2.11
Underlay approach considered in Zhang and Liang (2008).

It is also assumed that all users (CRs and PUs) are equipped with multiple antennas and that the CR transmitter exactly knows the channel matrices between itself and all the PUs $\{\mathbf{G}_k\}_{k=1}^K$. Palomar et al. (2003) considered the problem of designing the transmission covariance matrix $\mathbf{S} \triangleq E\big[\mathbf{x}[n]\mathbf{x}^H[n]\big] \in \mathbb{C}^{M_{t,s} \times M_{t,s}}$ that maximizes the instantaneous capacity of the CR pair under two constraints: the transmission power constraint and the interference power constraint. The CR's transmission power should not exceed P_t watts, while the interference should not exceed Γ_k at the kth PU. This problem was shown to encompass the design of optimal beamforming vectors and spatial multiplexing vectors in some special cases. In fact, it was shown that this problem can only be solved numerically using convex optimization techniques such as the interior point method (Boyd and Vandenberghe 2004). However, some special cases were considered where suboptimal solutions based on the singular value decomposition of the CR's channel matrix $\mathbf{H}$ were proposed.

Key to the solution of this optimization problem was assuming prior knowledge about the channel status between the CR and the PUs. In fact, this assumption is impractical since there is no direct communication between the two users. A similar approach was also studied in Zhang, Liang, and Xin (2008) in a different scenario; however, the principles are the same.

It is the difficulty in obtaining the desired information in the underlay approach that makes designing efficient MIMO systems a big challenge. However, this is not the case in the overlay approach. In spectrum overlay, CRs access the spectrum bands like regular users as long as the bands are empty.

Hence, the interference limit constraint faced in the underlay approach is not imposed herein. Obviously, this makes conventional design approaches like those in Palomar, Cioffi, and Lagunas (2003) directly applicable to the overlay approach. This is one of the reasons that made the research in the overlay approach focused more on employing cooperative access schemes rather than MIMO systems.

2.4.2 Cooperative Spectrum Access

As CRs cooperate with each other to sense the spectrum, they can also cooperate to access the spectrum. The idea of cooperative spectrum access is shown in Figure 2.12.

In this figure, a group of CRs collaborate to relay the message of the first CR to the base station. Since the CR network is covering a wide geographic area, each CR experiences different spectrum conditions and thus detects different spectrum holes. For instance, CR_1 in Figure 2.12 is only near PU_1, which uses f_1 for transmission; thus it has f_2, f_3, f_4 as available spectrum bands. Consequently, the CR network as a whole has a wealth of spectrum opportunities to be used for seamless transmission. To benefit from this setup, Ben Letaief and Zhang (2009) introduced the concepts of cognitive relay networks and cognitive space-time-frequency (STF) coding. We next briefly discuss each of these concepts.

Consider a CR network consisting of a source node intending to send a message to a destination node and a total of K relay nodes are available to aid this transmission process. Let us also assume that there is a group of PUs operating over a wide range of the spectrum. Let each CR fall within the transmission range of at least one PU. Furthermore, assume that the PU network is using a bank of N nonoverlapping frequency bands $f_1, f_2, \ldots, f_N$

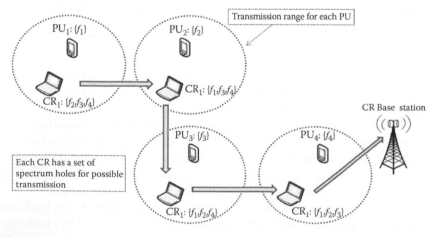

FIGURE 2.12
Cooperative CR network.

and that each PU can use all or part of these bands when operational (like an OFDMA-based system). Consequently, each CR gets a map of the available spectrum bands within its range using some sensing method. Hence, for the whole CR network, a spectrum map (matrix **B**) is composed as follows:

$$
\mathbf{B} = \begin{bmatrix} b_{1,1} & b_{1,2} & \cdots & b_{1,N} \\ b_{2,1} & b_{2,2} & \cdots & b_{2,N} \\ \vdots & \vdots & \ddots & \vdots \\ b_{K,1} & b_{K,2} & \cdots & b_{K,N} \end{bmatrix}_{K \times N}
\tag{2.11}
$$

where the $\{b_{k,n}\}_{k=1,n=1}^{K,N} = \{0,1\}$ is an indicator for the availability of the nth spectrum band to the kth relay node. For instance, the spectrum map matrix for the setup shown in Figure 2.12, ($K = N = 4$), is given by:

$$
\mathbf{B} = \begin{bmatrix} 0 & 1 & 1 & 1 \\ 1 & 0 & 1 & 1 \\ 1 & 1 & 0 & 1 \\ 1 & 1 & 1 & 0 \end{bmatrix}
\tag{2.12}
$$

Notice that for a spectrum band $\{f_n\}_{n=1}^N$ to be used by the kth CR, it needs to be available for the $\{k-1\}$th and $\{k+1\}$th CRs to avoid interference to the neighboring PUs.

The main advantage of cognitive relay networks is achieving seamless transmission if there is a continuous set of available spectrum bands between the source and the destination. However, it was shown in Ben Letaief and Zhang (2009) that if each relay node receives in one time slot and transmits in the next, the achievable data rate will be reduced significantly. In fact, it was shown that the larger the number of the relay nodes, the lower the achievable data rate. To overcome this degradation, Ben Letaief and Zhang (2009) proposed cognitive STF coding.

In the cognitive relay networks, each relay node uses at most two of the available spectrum bands to receive and then relay the message one step to the destination. By doing so, the CR network is in fact wasting a large portion of the available spectrum bands. Cognitive STF coding is intended to overcome this by fully exploiting the spectrum opportunities at each relay node in the network. The proposed scheme works as follows:

1. In the first phase, the source broadcasts a block of N_s symbols in N_s time slots to all intermediate relay nodes.

2. All relay nodes then decode the received signal and re-encode it according to some coding scheme before forwarding it to the destination.

It was analytically shown that any conventional orthogonal space-time block coding (STBC) can be used as long as all nodes use all available frequencies to relay the message to the destination. Furthermore, it was also shown that the achievable rate does not depend on the number of relay nodes.

2.5 Conclusion

This chapter focused on the cooperative dynamic spectrum access in CR networks. This research area is a new extension of the emerging CR technology. It aims at bringing the advantages of MIMO systems and cooperative communications to the CR technology. Dynamic spectrum access is a core task lying at the heart of CR technology. It is the enabling parameter that allows CRs to sense the spectrum holes prior to accessing them for opportunistic transmission. It is a two-stage process: First, a spectrum sensing method is employed to detect the spectrum holes, and then the CR chooses the best way to exploit this hole to transmit data. In this chapter, we emphasized utilizing MIMO systems and the emerging cooperative communications to enhance the dynamic spectrum access process.

References

Arslan, H. (Ed.). (2007). *Cognitive Radio, Software Defined Radio, and Adaptive Wireless Systems*. Dordrecht, Netherlands: Springer.

Ben Letaief, K., and Zhang, W. (2009, May). Cooperative communications for cognitive radio networks. *Proceedings of the IEEE* 97(5), 878–893.

Boyd, S., and Vandenberghe, L. (2004). *Convex Optimization*. Cambridge: Cambridge University Press.

Cabric, D., Mishra, S.M., and Brodersen, R.W. (2004, November). Implementation issues in spectrum sensing for cognitive radios. *Proceedings of the 38th Asilomar Conference on Signals, Systems and Computers*.

Cabric, D., Mishra, S.M., Willkomm, D., Brodersen, R.W., and Wolisz, A. (2005, June). A cognitive radio approach for usage of virtual unlicensed spectrum. *Proceedings of the 14th IST Mobile and Wireless Communications Summit*.

Cave, M., Doyle, C., and Webb, W. (2007). *Essentials of Modern Spectrum Management*. Cambridge: Cambridge University Press.

Clancy, T.C. (2007, November). Formalizing the interference temperature model. *Wireless Communications and Mobile Computing* 7(9), 1077–1086.

Digham, F., Alouini, M.-S., and Simon, M.K. (2003). On the energy detection of unknown signals over fading channels. *Proceedings of the IEEE International Conference on Communications (ICC)*, Anchorage, Alaska, pp. 3575–3579.

Digham, F., Alouini, M.-S., and Simon, M.K. (2007, January). On energy detection of unknown signals over fading channels. *IEEE Transactions on Communications* 55, 21–24.

Establishment of Interference Temperature Metric to Quantify and Manage Interference and to Expand Available Unlicensed Operation in Certain Fixed Mobile And Satellite Frequency Bands. (2003). Washington, D.C.: Federal Communications Commission (FCC).

Fitzek, F.H., and Katz, M.D. (Eds.). (2007). *Cognitive Wireless Networks, Concepts, Methodologies and Visions Inspiring the Age of Enlightenment of Wireless Communications.* Dordrecht, Netherlands: Springer.

Ganesan, G., and Li, Y.G. (2007a, June). Cooperative spectrum sensing in cognitive radio, Part I: Two user networks. *IEEE Transactions on Wireless Communications* 6(6), 2204–2213.

Ganesan, G., and Li, Y.G. (2007b, June). Cooperative spectrum sensing in cognitive radio, Part II: Multiuser networks. *IEEE Transactions on Wireless Communications* 6(6), 2214–2222.

Ghasemi, A., and Sousa, E.S. (2005). Collaborative spectrum sensing for opportunistic access in fading environments. *Proceedings of the IEEE Symposium on New Frontiers in Dynamic Spectrum Access Networks (DySPAN)*, Baltimore, Md., pp. 131–136.

Han, N., Shon, S., Chung, J.H., and Kim, J.M. (2006). Spectral correlation based signal detection method for spectrum sensing in IEEE 802.22 WRAN systems. *Proceedings of the 8th International Conference on Advanced Communication Technology (ICACT)*, Phoenix Park, Korea, pp. 1765–1770.

Han, Z., Fan, R., and Jiang, H. (2009, June). Replacement of spectrum sensing in cognitive radio. *IEEE Transactions on Wireless Communications* 8(6), 2819–2826.

Haykin, S. (2005, February). Cognitive radio: Brain-empowered wireless communications. *IEEE Journal on Selected Areas in Communications* 23(2), 201–220.

Herath, S.P., and Rajatheva, N. (2008). Analysis of equal gain combining in energy detection for cognitive radio over Nakagami channels. *Proceedings of the IEEE Global Telecommunications Conference (GLOBECOM)*, New Orleans, La., pp. 1–5.

Herath, S.P., Rajatheva, N., and Tellambura, C. (2009). Unified approach for energy detection of unknown deterministic signal in cognitive radio over fading channel. *Proceedings of the IEEE International Conference on Communications (ICC)*, Dresden, Germany.

Hulbert, A.P. (2005). Spectrum sharing through beacons. *Proceedings of the 16th Annual IEEE International Symposium on Personal Indoor and Mobile Radio Communications (PIMRC)*, Berlin, Germany, pp. 11–14.

Hur, Y., Park, J., Woo, W., Lee, J.S., Lim, K., Lee, C.-H., et al. (2006). WLC05-1: A cognitive radio (CR) system employing a dual-stage spectrum sensing technique: A multi-resolution spectrum sensing (MRSS) and a temporal signature detection (TSD) technique. *Proceedings of the IEEE Global Telecommunications Conference (GLOBECOM)*, San Francisco, pp. 1–5.

Kostylev, V.I. (2002). Energy detection of a signal with random amplitude. *Proceedings of the IEEE International Conference on Communications (ICC)*, New York, pp. 1606–1610.

Lai, L., V. Fan, and V. Poor. (2008). Quickest detection in cognitive radio: A sequential change detection framework. *Proceedings of the IEEE Global Telecommunications Conference (GLOBECOM)*, New Orleans, La.

Larsson, E.G., and Stoica, P. (2003). *Space-Time Block Coding for Wireless Communications*. Cambridge: Cambridge University Press.

Li, H., Li, C., and Dai, H. (2008). Quickest spectrum sensing in cognitive radio. *Proceedings of the 44th Annual Conference on Information Sciences and Systems (CISS)*, Princeton, N.J., pp. 203–208.

Lundén, J., Koivunen, V., Huttunen, A., and Poor, H. V. (2009, November). Spectrum sensing in cognitive radios based on multiple cyclic frequencies. *IEEE Transactions on Signal Processing* 57(11), 4182–4195.

Luo, L., Neihart, N.M., Roy, S., and Allstot, D.J. (2009, June). A two-stage sensing technique for dynamic spectrum access. *IEEE Transactions on Wireless Communications* 8(6), 3028–3037.

Ma, J., Li, G., and Juang, B. (2009, May). Signal processing in cognitive radio. *Proceedings of the IEEE* 97(5), 805–823.

Mangold, S., Zhong, Z., Challapali, K., and Chou, C.-T. (2004). Spectrum agile radio: Radio resource measurements for opportunistic spectrum usage. *Proceedings of the IEEE Global Telecommunications Conference (GLOBECOM)*, Dallas, pp. 3467–3471.

Mishra, S.M., Sahai, A., and Brodersen, R.W. (2006). Cooperative sensing among cognitive radios. *Proceedings of the IEEE International Conference on Communications (ICC)*, Istanbul, Turkey, pp. 1658–1663.

Mitola III, J. (2000). *Cognitive radio: An integrated agent architecture for software defined radio*. PhD diss., Royal Institute of Technology (KTH), Stockholm.

Notice of Proposed Rule Making and Order. (2003). Washington, D.C.: Federal Communications Commission (FCC).

Oner, M., and Jondral, F. (2007, April). On the extraction of the channel allocation information in spectrum pooling systems. *IEEE Journal on Selected Areas in Communications* 25(3), 558–565.

Palomar, D.P., Cioffi, J.M., and Lagunas, M.A. (2003, September). Joint Tx-Rx beamforming design for multicarrier MIMO channels: A unified framework for convex optimization. *IEEE Transactions on Signal Processing* 51(9), 2381–2401.

Principles for Promoting the Efficient Use of Spectrum by Encouraging the Development of Secondary Markets. (2003). Washington, D.C.: Federal Communications Commission (FCC).

Quan, Z., Cui, S., Poor, V.H., and Sayed, A. (2008, November). Collaborative wideband sensing for cognitive radio. *IEEE Signal Processing Magazine* 25(6), 60–73.

Re: Establishment of an Interference Temperature Metric to Quantify and Manage Interference and to Expand Available Unlicensed Operation in Certain Fixed, Mobile, and Satellite Frequency Bands. (2007, May). Washington, D.C.: Federal Communication Commission.

Sahai, A., Hoven, N., and Tandra, R. (2004). Some fundamental limits on cognitive radio. *Proceedings of the Allerton Annual Conference on Communication, Control, and Computing*. Monticello, Ill.

Sendonaris, A., Erkip, E., and Aazhang, B. (2003a, November). User cooperation diversity—Part I: System description. *IEEE Transactions on Communications* 51(11), 1927–1938.

Sendonaris, A., Erkip, E., and Aazhang, B. (2003b, November). User cooperation diversity—Part II: Implementation aspects and performance analysis. *IEEE Transactions on Communications* 51(11), 1939–1948.

Spectrum Policy Task Force Report. (2002). Washington, D.C.: Federal Communications Commission (FCC).

Tang, H. (2005). Some physical layer issues of wide-band cognitive radio systems. *Proceedings of the IEEE Symposium on New Frontiers in Dynamic Spectrum Access Networks (DySPAN)*, Baltimore, Md., pp. 151–159.

Tian, Z., and Giannakis, G.B. (2006). A wavelet approach to wideband spectrum sensing for cognitive radios. *Proceedings of the International Conference on Cognitive Radio Oriented Wireless Networks and Communications (CROWNCOM*, Mykonos Island, Greece, pp. 1–5.

Wild, B., and Ramchandran, K. (2005). Detecting primary receivers for cognitive radio applications. *Proceedings of the IEEE Symposium on New Frontiers in Dynamic Spectrum Access Networks (DySPAN)*, Baltimore, Md., pp. 124–130.

Xiao, Y., and Hu, F. (Eds.). (2009). *Cognitive Radio Networks.* Boca Raton, Fla.: CRC Press.

Yucek, T., and Arslan, H. (2009). A survey of spectrum sensing algorithms for cognitive radio applications. *IEEE Communications Surveys and Tutorials* 11(1), 116–130.

Zeng, Y., and Liang, Y.-C. (2007). Covariance based signal detection for cognitive radio. *Proceedings of the IEEE Symposium on New Frontiers in Dynamic Spectrum Access Networks (DySPAN)*, Dublin, Ireland, pp. 202–207.

Zeng, Y., and Liang, Y.-C. (2009, June). Eigenvalue-based spectrum sensing algorithms for cognitive radio. *IEEE Transactions on Communications* 57(6), 1784–1793.

Zhang, L., Liang, Y.-C., and Xin, Y. (2008, January). Joint beamforming and power allocation for multiple access channels in cognitive radio networks. *IEEE Journal on Selected Areas in Communications* 26(1), 38–51.

Zhang, R., and Liang, Y.-C. (2008, February). Exploiting multi-antennas for opportunistic spectrum sharing in cognitive radio networks. *IEEE Journal of Selected Topics in Signal Processing* 2(1), 88–102.

Zhao, Q., and Sadler, B.M. (2007, May). A survey of dynamic spectrum access. *IEEE Signal Processing Magazine* 24(3), 79–89.

3

Adaptive Modulation, Adaptive Power Allocation, and Adaptive Medium Access

3.1 Introduction

The inherent requirements for wireless sensor networks (WSNs) to work under complex conditions introduce a substantial number of constraints [1] [2]. A few essential issues that still challenge the research community include:

- *Realistic Protocol Design:* Many of the current WSN platforms are developed with assumptions that simplify the wireless communication process and the operation environments [3]. The lack of realistic models for system design often makes the solutions work well in simulation but not robust enough for actual networks. Research is needed to focus on developing better models and new network protocols for the realistic sensing environments.

- *Power Management:* Due to the very limited capacity of the battery powering the sensor nodes, energy is a precious resource in the network. The fact that most sensor network applications require a long operating lifetime emphasizes the importance of research to improve energy efficiency in WSNs [4].

- *Real-Time Operation:* WSNs are highly time constrained. In many of the applications, sensing information needs to be collected within a short time frame in order to make the data acquisition valid and accurate. However, most current protocols do not meet the real-time operation requirement satisfactorily. This leads to the need for designing real-time operation protocols that can sufficiently reduce delays [5].

- *Security and Privacy Issues:* Sensor nodes are normally deployed over large and accessible areas. Unencrypted information may be intercepted during transmission. To ensure privacy in the system, security issues must be considered and properly addressed in every component.

Notwithstanding, WSNs are gaining increasing popularity due to many attractive features in flexibility, cost-efficiency, high resolution, cooperative effort, and self-organizing capabilities [1] [4] [6]. Although each node is only capable of a limited amount of processing, the coordination of a large group of sensor nodes can form a WSN able to sense the environment in great detail [7].

The main results of this chapter were originally derived in [65] and [66] and are summarized below:

1. Different link adaptation policies are evaluated for energy saving. The goal of this analysis is to achieve optimal spectral efficiency while minimizing energy consumption in the network, thus extending the network operating lifetime while simultaneously meeting the quality of service (QoS) requirements. The work of Goldsmith et al. [8] is extended to compute the energy performance in the network where the total available energy constraints are imposed on all nodes in the communication path. Data rate and transmit power, the two key factors determining energy consumption in the network, are studied.

2. An adaptive sleep with adaptive modulation (ASAM) algorithm is used to adaptively adjust the durations of the node operating stages in the wireless channels for minimizing energy expenditure and enhancing the network lifetime. Relevant formulas for calculating energy consumption are derived for different medium access control (MAC) layer protocols. The energy consumption of adaptive modulation (AM), adaptive modulation with idle mode (AMI) and the ASAM algorithm is evaluated and compared.

3. Adaptive power control and allocation algorithms are introduced to analyze the overall achievable rate and power level in single-hop and multihop communications, respectively. Optimal power control and allocation factors are also derived. An example of a two-link multihop network is explored using different adaptive transmission protocols.

This chapter is organized as follows. Section 3.2 provides the necessary background and analytical framework for this work. Relevant parameters and system models are described. Section 3.3 presents the literature review of link adaptation and feedback communication systems. Recent research and designs for energy-saving protocols are reviewed. The fundamentals of different adaptive techniques are also explained. Section 3.4 describes the ASAM algorithm and adaptive power allocation policy. Section 3.5 presents the simulation results. The effects of channel conditions, traffic intensity, and number of modulation stages are deeply investigated. The per-node operating lifetime for point-to-point communication and multihop networks using the link adaptation polices and ASAM is evaluated and discussed.

3.2 System Model

In this section we are particularly interested in physical (PHY) layer channel characteristics. Figure 3.1 displays the architecture of a typical wireless communication system and the main components of interest.

3.2.1 Information Source and Sink

The communication information data $x(t)$ are generated by the source unit. In order to reduce the implementation complexity in practice, it is assumed that the data are uniformly distributed and are generated at a fixed symbol rate (see Section 3.3.5 for details). The communication system bandwidth is denoted as B, and R_s is the fixed symbol rate. The data are transmitted through a flat-fading wireless channel where additive white Gaussian noise (AWGN) is added during the process.

3.2.2 Transmitter

The transmitter in this model has three main tasks:

- *Transmission:* The transmitter sends the data through a wireless channel to a receiver based on the selected modulation schemes and power levels.
- *Adaptive Modulation:* Unlike in a traditional modulator, the modulation scheme here is not fixed but varies based on QoS criteria. In this section, an instantaneous bit-error-rate (BER) requirement is further

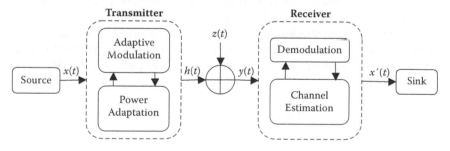

FIGURE 3.1
System architecture for a typical wireless communication system.

imposed. Therefore, the data rates and the transmission durations are determined by different modulation schemes, which in turn are selected according to the performance requirements.

- *Power Adaptation:* The most appropriate power adaptation policy (PAP) is selected to optimize power levels according to the channel state information (CSI) feedback from the receiver. Two techniques of power adaptation are considered in this chapter—namely, power allocation and adaptive sleep—both discussed in detail in Section 3.4.

3.2.3 Receiver

The receiver is responsible for receiving data, demodulating the received information, and estimating the CSI after the wireless channel. Hence a feedback channel is further required so that the estimated CSI can be fed back to the transmitter. Normally there might be estimation errors and delays involved in this feedback process, which negatively affects the transmitter decisions. However, we assume that the channel estimation is perfect for the receiver and the delay from the feedback channel is negligible.

3.2.4 Wireless Channel

During the processes of transmission, reception, and signal propagation, the additive noise should be considered in the analysis. Considering the channel gain and additive noise, the output signal can be expressed as:

$$y = h \cdot x + z \tag{3.1}$$

where x is the input signal, h is the channel gain, and z is the AWGN, here formed by taking a zero-mean white Gaussian random variable with power spectral density (PSD) of $N_0/2$.

3.2.5 Lognormal Shadowing Channel Model

Wireless links experience shadowing and fading effects. In this section, the shadowing effect of the channel is modeled as a lognormal distribution that describes the random shadowing effects in the link budget calculation. The received power level is formulated by considering the lognormal distributed path loss. Using this approximation, a simple path-loss model can be expressed as [9]:

$$P_r = P_0 - 10n \log_{10}\left(\frac{d}{d_0}\right) + X_\sigma \tag{3.2}$$

where P_r is the received power and P_0 is the received power at the reference distance d_0 from the transmitter; d is the distance between the transmitter and the receiver; X_σ is a zero-mean Gaussian distributed random variable with standard deviation σ; and n is the path-loss exponent that is dependent on the propagation environment. The typical values of n are in the range of [2, 5] for common wireless sensor environments [10].

3.2.6 Rician Fading Channel Model

Here the multipath fading effects are modeled by Rician distribution.

When analyzing the fading probability distribution of the wireless channel, Rician PDF is normally represented as a function of the received signal to noise ratio (SNR) (γ) [10]:

$$P_{\Gamma}(\gamma) = \frac{(1+K)}{\bar{\gamma}} e^{\left(-K-\frac{(1+K)\gamma}{\bar{\gamma}}\right)} I_0\left(2\sqrt{\frac{K(1+K)\gamma}{\bar{\gamma}}}\right), \gamma \geq 0. \tag{3.3}$$

where I_0 is the modified Bessel function of the first kind with order zero, and K is the Rician K-factor defined as the ratio of dominant component signal power over the local mean scattered power. In other words, the K-factor shows the strength of the line-of-sight (LOS) signal in the channel [11]. In this section, links are uncorrelated so that the shadowing and fading effect on one link has no impact on others. Thus, the channel Rician K-factors can be varied independently for each link.

3.3 Adaptive Transmission and Feedback Communication System

3.3.1 Introduction

Adaptive modulation (AM) or link adaptation allows for dynamic adjustment of the communication systems (for example, by changing modulation [12] and coding [13] schemes and other parameters [14]), according to the characteristics of the time-varying channel [15]. This effectively improves spectral efficiency by adapting transmission parameters to specific communication conditions.

The basic concept behind adaptive transmission is to maintain a stable level of E_b/N_0 by varying the transmit power, data rate, coding rate, coding scheme, or a combination of them. It takes advantage of favorable channel conditions by transmitting data at high speed. In this way, spectral efficiency is improved while maintaining BER requirements [15].

3.3.2 Adaptive System Design

Figure 3.2 illustrates the system model for a typical adaptive transmission system. Unlike in traditional communication systems, modulation level and

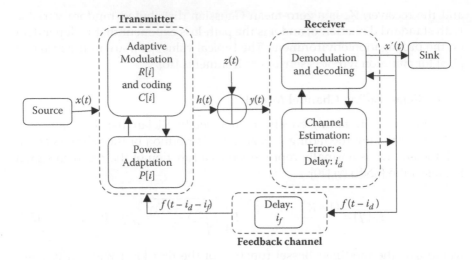

FIGURE 3.2
Adaptive feedback system model.

transmit power in adaptive systems are not fixed but dynamically controlled by transmitter. Through the feedback channel, the transmitter exchanges information with the receiver and collects the current CSI data. The transmitter can then make decisions on the proper transmission parameters to use. During the process, both adaptive modulation control and power adaptation units are functioning simultaneously to ensure modulation schemes and transmit power are selected accurately.

The receiver has two functionalities: demodulation and estimation. It cooperates with the transmitter to determine the transmitted information and to estimate the CSI. During the estimation and feedback process, delay and error can occur, impairing the accuracy of the estimates.

3.3.3 Link Adaptations

The parameters for adjustment can include: symbol rate [16], modulation schemes or constellation size, transmit power [17], data transmission rate [12] [18] [19] [20], and coding parameters [21] [22] [23] [24]. These parameters can be varied either individually or jointly according to application and performance requirements. Recent researches have focused on adapting one or two parameters, specifically, power and/or rate [15] [16] [20] [19] [25], rate and coding [13] [21], and BER [21].

The work by Goldsmith et al. [20] indicates that the Shannon capacity of a flat-fading channel can be best achieved by jointly varying the data rate and transmit power. In addition, since the data rate and the transmit power are two parameters for calculating the network energy, variable-rate/variable-power adaptation becomes the most interesting link adaptation policy in research when analyzing the network energy consumption. In this section, the link adaptation system performance is evaluated using variable rate and variable power both individually and jointly. The objective is to solve the energy optimization problem so that the maximum node lifetime is achieved under network constraints.

Early studies on adaptation techniques were done by Cavers [16] and Hayes [17] in the 1960s. The results were promising for adaptive techniques to improve the system efficiency and throughput by supporting various communication profiles and multiple transmission rates according to the link quality. However, due to the hardware limitations and difficulties in accurate estimation of the CSI, link adaptation was considered unfeasible at that time [8]. With advances in technology, especially in hardware, the original concerns of link adaptation have become less prohibitive. Technologies that incorporate AM into wireless local area networks (WLANs) [26] [27] [28] [29], multiple-input/multiple-output (MIMO) communications [30], and 3G/4G cellular networks have revived [25] [31] [32] [33] [34].

It has to be noted that the adaptive systems must be designed to ensure that consistent communication requirements are met. Link adaptation protocols can avoid the poor utilization of the channel even with deep fading. Therefore, link adaptation increases the spectral efficiency by transmitting at desirable speed based on channel conditions. Particularly for multihop relay networks, communication links can transmit at different data rates and power levels to achieve optimal spectral efficiency and minimum energy consumption.

However, link adaptation still suffers from a few practical limitations [35]. First, it is highly dependent on reliable receivers and feedback channels to estimate and relay accurate CSI to transmitters. In addition, the communication process requires real-time estimation and error-free transmission in order to ensure the accuracy of the transmitter decisions. If the time-varying CSI cannot be received correctly or timely, link adaptation is not possible. Furthermore, most link adaptation techniques are evaluated mainly in terms of simulations. Actual experiments in realistic networks are desperately needed.

3.3.4 Link Adaptation for Energy-Constrained Networks

Applying link adaptation to energy-constrained sensor networks has received a significant amount of interests in research [36] [38] [39] [40]. Many designs offer optimal energy performance by using extra low-power components in sensor nodes [37] [41] [42]. However, additional hardware can introduce noise and errors to the network and influence the system performance [35]. Some works also consider using dynamic routing protocols for delivering the information from source to destination [43] [44] [45]. The design often focuses on balancing battery power for all nodes along the transmission routes.

Recent research in the field of energy-constrained sensor networks also proposes the concept of energy-aware protocols to control the power levels during transmission [46–49, 64]. Instead of being fixed beforehand, the transmit power is dynamically allotted. Energy-aware protocols balance the link budget through adaptive variation, so that each transmission can spend a different amount of energy given the channel conditions [17]. Thus, without wasting the power or sacrificing BER, the protocol provides high average link energy efficiency by taking advantage of flat-fading through adaptation. No energy is wasted during the process, hence maximizing the node lifetime.

One of the essential ideas in these energy-aware protocols is to power down the nodes when they are not performing any tasks. This is due to the fact that in WSNs, nodes normally need to be communicating for a short period of time only; thus a high degree of redundancy usually exists in network topology [30]. It is possible to design wireless communication protocols to minimize energy consumption by letting the nodes sleep for the maximum amount of time. Adaptive sleep (AS) technique, therefore, has been proposed so that node sleep time can be adjusted based on current fading conditions. Most of the time when there is no communication occurring, nodes are powered down and operated at the minimum power level. This standby power can be orders of magnitude lower than the active power. Energy efficiency is thereby improved as sleep time is maximized in the network.

3.3.5 Adaptive Techniques

Dynamically adjusting the transmission parameters leads to various adaptive techniques. This chapter investigates the network energy consumption by considering data rate and transmit power. For a given BER constraint, the spectral efficiency has to be optimized, which applies to all adaptive techniques. The approximations and formulas for different link adaptation policies are presented in the following sections.

3.3.5.1 BER Approximation for Multiple Quadrature Amplitude Modulation

The spectral efficiency is the main advantage to be gained from AM analysis. It is defined as the average data rate per unit bandwidth (R/B). The transmission rates are determined by the modulation schemes, that is, $r(\gamma) = \log_2[M(\gamma)]$ (bits/symbol) [8]. Therefore, the spectral efficiency for continuous rate adaptation [Equation (3.4)] and the discrete rate adaptation [Equation (3.5)], respectively, are given by [8]:

$$\frac{R}{B} = \int_{0}^{\infty} r(\gamma)p(\gamma)d\gamma \tag{3.4}$$

$$\frac{R}{B} = \sum_{i=0}^{N-1} r_i \int_{\gamma_i}^{\gamma_{i+1}} p(\gamma)d\gamma \tag{3.5}$$

The adaptive rate $r(\gamma)$ is determined by the modulation schemes typically restricted by an average transmit power $\bar{P}$. The transmit power constraint in this assumption is given by:

$$\int_0^\infty P(\gamma)p(\gamma)d\gamma \le \bar{P} \qquad (3.6)$$

With Gray bit mapping, the expression for the BER of multiple quadrature amplitude modulation (MQAM) can be approximated as a function of the receiver SNR

$$\gamma\left(\frac{P(\gamma)}{\bar{P}}\right)$$

and the constellation size M [8]:

$$BER_{MQAM}(\gamma) \approx \frac{2}{\log_2 M}\left(1-\frac{1}{\sqrt{M}}\right) \times erfc\left(\sqrt{1.5\frac{\gamma\frac{P(\gamma)}{\bar{P}}}{M-1}}\right) \qquad (3.7)$$

As this expression cannot be easily solved for its power $P(\gamma)$ and its rate $R = \log_2 M$, it can be considered as a tight approximation of the following equation for constellation size $M \ge 2$ and BER less than 10^{-3} [8]:

$$BER_{MQAM}(\gamma) \le 0.2\exp\left(\frac{-1.5\gamma\frac{P(\gamma)}{\bar{P}}}{M-1}\right) \qquad (3.8)$$

This tight approximation for rate variation and power adaptation is used in this chapter for network lifetime analysis.

3.3.5.2 Variable Rate

Considering the data transmission rate varies with channel gain, the network can achieve its optimal transmission rate by two means: (1) fixing the symbol rate and using multiple modulation schemes or constellation sizes, or (2) fixing the modulation scheme and changing the symbol rate. Usually, the second method is difficult to implement because varying the symbol rate requires varying the signal bandwidth, which is complicated in practice [23]. In contrast, changing the modulation types or the constellation sizes is more feasible. For this reason, it is employed in this chapter;

that is, the modulation schemes are varied to achieve the optimal rate during transmission.

To formulate the variable-rate modulation problem, we consider a family of MQAM signals with a fixed symbol rate T_s, with M denoting the constellation size of the modulation scheme, $\bar{P}$ denoting the average transmit power, and N_0 and B being the noise and bandwidth, respectively. Assuming ideal Nyquist pulses for each constellation and unity average channel gain, the average received SNR is expressed as [8]:

$$\bar{\gamma} = \frac{\bar{P}T_s}{N_0} = \frac{\bar{P}}{N_0 B} = \frac{E_s}{N_0} \qquad (3.9)$$

Recall that the spectral efficiency is defined as data rate per unit bandwidth R/B. For a fixed M, it becomes $R/B = (\log_2 M)/B$ bits/symbol. In this case, we are only considering the variable rate. The spectral efficiency is therefore parametrized by the average transmit power and the BER. Rearrange Equation (3.9) in terms of M, the expression for the constellation size is obtained as a function of the received SNR γ:

$$M(\gamma) = 1 + \frac{-1.5\gamma}{\ln(5BER)} \frac{P(\gamma)}{\bar{P}} \qquad (3.10)$$

The relation between symbol rates and received SNRs for continuous and discrete rate adaptation is depicted in Figure 3.3. The spectral efficiency can then be optimized by maximizing:

$$E[r(\gamma)] = E\left[\log_2\left(M(\gamma)\right)\right] = \int \log_2\left(1 + \frac{-1.5\gamma}{\ln(5BER)} \frac{P(\gamma)}{\bar{P}}\right) p(\gamma) d\gamma \qquad (3.11)$$

with respect to the power constraint in Equation (3.6).

3.3.5.3 Variable Power

The transmit power can be adapted to compensate for SNRs. The goal is to maintain a fixed BER and, equivalently, a constant received SNR. In [8], two techniques are proposed for fixed-rate variable-power adaptation: channel inversion adaptation and truncated channel inversion. Although both techniques aim to maintain a constant received SNR, the former suffers from a larger power penalty since most of the average signal power is used to compensate for deep fading condition, while the latter provides a cutoff level below which no signal is transmitted. The power adaptation formula for channel inversion is

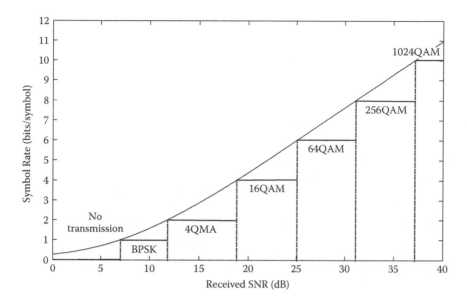

FIGURE 3.3
Symbol rate verification using adaptive modulation when BER = 10^{-4}.

$$\frac{P(\gamma)}{\overline{P}} = \frac{\sigma}{\gamma} \tag{3.12}$$

where σ is the constant received SNR, and γ is the channel gain.

Using this power adaptation policy, the spectral efficiency is obtained by substituting Equation (3.12) into (3.11):

$$\frac{R}{B} = \log_2\left(1 + \frac{-1.5\gamma}{\ln(5BER)E[1/\gamma]}\right) \tag{3.13}$$

When using the truncated channel inversion for power adaptation, the fading can be inverted above a given cutoff γ_0. The power adaptation is [8]:

$$\frac{P(\gamma)}{\overline{P}} = \begin{cases} \frac{\sigma}{\gamma} & \gamma \geq \gamma_0 \\ 0 & \gamma \leq \gamma_0 \end{cases} \tag{3.14}$$

Similarly, the spectral efficiency for truncated channel inversion is given by:

$$\frac{R}{B} = max_{\gamma 0} \log_2\left(1 + \frac{-1.5\gamma}{\ln(5BER)E_{\gamma 0}[1/\gamma]}\right) \tag{3.15}$$

By introducing a power control technique to determine the power level needed for a successful transmission, power control values are formulated by rearranging Equation (3.10) in terms of $P(\gamma)/\bar{P}$. We then obtain the expression of the normalized power control factor as:

$$\frac{P(\gamma)}{\bar{P}} = \frac{\ln(5BER) \times (M-1)}{-1.5 \cdot \gamma} \tag{3.16}$$

For each received SNR value, the instantaneous power value $P(\gamma)$ can be calculated as the product of the power control factor $P(\gamma)/\bar{P}$ and the average power $\bar{P}$. Thus, instead of using a fixed average power for transmission at all times, the instantaneous power level can be varied according to the channel conditions. The power control factors relative to the received SNRs are displayed in Figure 3.4. As shown in the figure, the power level is reduced with increased SNR for the same modulation scheme. Significant power increase occurs only when the transmitter is switching from the lower modulation levels to the higher ones.

3.3.5.4 Adaptive Rate and Power for MQAM Modulation Scheme

By combining rate adaptation and variable-power techniques, the spectral efficiency can be further improved. Optimal spectral efficiency can be determined for four cases: continuous rate adaptation with an average BER constraint (C-Rate A-BER), continuous rate adaptation with an instantaneous BER constraint (C-Rate I-BER), discrete rate adaptation with an average BER

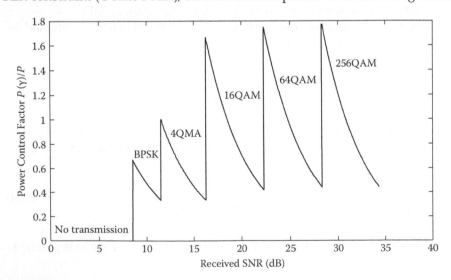

FIGURE 3.4
Normalized power allocation for MQAM.

constraint (D-Rate A-BER), and discrete rate adaptation with an instanta-neous BER constraint (D-Rate I-BER) [18].

In this section, the discrete rate continuous power adaptation (see Section 3.4.2 for details) is used to maximize the spectral efficiency with the con-straints of average power and average BER. This then becomes a constrained optimization problem that can be solved using the Lagrange method. The general Lagrange equation for this optimization problem subject to the power and BER constraints is given by:

$$f_{opt} = f_r(\gamma) + \lambda_1 \sum_{i=1}^{n} BER_i(\gamma) + \lambda_2 \sum_{i=1}^{m} P_i(\gamma) \tag{3.17}$$

where f_r is the energy function; BER_i is the instantaneous BER constraints; and P_i is the average transmit power constraints subject to each SNR value. The optimal rate and power can be satisfied by solving the following equation:

$$\frac{\partial f_{opt}}{\partial r(\gamma)} = 0 \text{ and } \frac{\partial f_{opt}}{\partial P(\gamma)} = 0 \tag{3.18}$$

where $r(\gamma)$ and $P(\gamma)$ are the rate and power, respectively. Both of them are non-negative for all SNR γ.

3.4 Multihop Relay Network and Energy-Constrained Network Analysis

3.4.1 Energy Consumption with Adaptation Techniques

The general formula for energy consumption can be expressed as:

$$E = P \times T \tag{3.18}$$

where P is the power level and T is the transmission time determined as the reciprocal of the data rate: $T = 1/R$. Therefore, energy consumption is also dependent on the communication data rate.

In this section, we consider four operation modes with different power levels for the sensor nodes:

- *Transmission mode:* The packet is transmitted from one node to another. Both radio transmission and the central processing unit

(CPU) are active in transmission mode. The device is able to activate the processor, listen to the channel, wait for reception, and transmit data. The transmission mode requires the highest power level compared to the other modes.

- *Active mode:* The node is awake and waiting for receiving the packets. In this mode, the transmission is off while the CPU maintains a high functionality. The current consumption now is referred to as "run current," which supports a high volume of CPU activities. The active mode spends less energy than the transmission mode.

- *Idle mode:* Here, the CPU maintains low activities. The radio transmission is off. Therefore, the current consumption in this mode is lower than in both the transmission and the active modes, but higher than the sleep mode.

- *Sleep mode:* After the data is successfully delivered, the node hibernates. Most node components are powered down and placed on standby to save energy. To wake up the transmitter and CPU again, it only needs a clock-type signal. The power consumption in this mode is the lowest among the four modes.

The energy consumed at each stage is determined by its power level and operating duration. This section studies the per-node lifetime during communication by considering the energy consumption of the four different operating stages. Link adaptation technique is analyzed in point-to-point communication for single-hop networks. In addition, adaptive power allocation factors for multihop relay networks are also derived.

3.4.2 Single-Hop Discrete Rate Continuous Power Adaptation

We first investigate the single-hop network case where the source directly transmits to its destination. Here, we consider the discrete rate continuous power adaptation with instantaneous BER adaptation [35]. For a given set of discrete rates $\{r_i\}_{i=1}^{N-1} = \log_2(M_i)$, each rate remains constant in the region $[\gamma_i, \gamma_{i+1})$, where $i = 1, 2, \ldots, N-1$. According to Equations (3.15) and (3.16), the optimal power adaptation is expressed as [8]:

$$\frac{P(\gamma)}{\bar{P}} = \frac{h(r_i)}{\gamma} \tag{3.19}$$

where

$$h(r_i) = \frac{ln(5BER)}{-1.5} \times (M_i - 1)$$

and $BER(\gamma) = \overline{BER}$ in this case.

The optimal rate region boundaries can be found via the Lagrange optimization method. The Lagrange equation for a discrete rate continuous power protocol is given by [18]:

$$J(\gamma_1, \gamma_2, \ldots, \gamma_N) = \sum_{0 \leq i \leq N-1} r_i \int_{\gamma_i}^{\gamma_{i+1}} p(\gamma) d\gamma + \lambda \left[\sum_{0 \leq i \leq N-1} \int_{\gamma_i}^{\gamma_{i+1}} \frac{h(r_i)}{\gamma} p(\gamma) d\gamma - 1 \right] \quad (3.20)$$

Solve the equations:

$$\frac{\partial J}{\partial \gamma_i} = 0, \quad 0 \leq i \leq N-1 \quad (3.21)$$

Therefore, the SNR thresholds of the optimal power can be found as:

$$\gamma_0 = \frac{h(r_0)}{r_0} \rho \quad (3.22)$$

and

$$\gamma_i = \frac{h(r_i) - h(r_{i-1})}{r_i - r_{i-1}} \rho, \quad 0 \leq i \leq N-1 \quad (3.23)$$

where ρ is a constant determined by the average power constraints in Equation (3.6).

Interpreting this equation in a discrete rate manner, the average power can be determined by:

$$\sum_{0 \leq i \leq N-1} \int_{\gamma_i}^{\gamma_{i+1}} \frac{h(r_i)}{\gamma} p(\gamma) d\gamma = 1 \quad (3.24)$$

Since source nodes directly transmit to the destination, energy consumption in the network can be quantified by the node lifetime. The node lifetime, in turn, is defined as the period of time during which the node has sufficient energy for transmitting information to its destination. The objective of single-hop discrete rate continuous power adaptation is then to achieve maximum node lifetime by adaptively changing the transmission rate and power level according to the channel conditions, based on the discrete rate threshold values derived via Equations (3.22)–(3.24).

3.4.3 Multihop Relay Networks

In the previous section, the node lifetime in single-hop WSNs was discussed. However, a typical WSN system usually consists of several nodes. These nodes communicate with each other and form a multihop network. In this scenario, we focus on the lifetime of the overall network instead of the lifetime of a single node.

In particular, we are concerned with two types of multihop networks: (1) relay networks in which source data are transmitted to the destination via relay nodes, and (2) multiple-link networks in which multiple wireless communication links are formed by several independent sources and destinations. In both cases, wireless channels can be easily influenced by each other. As a result, the shadowing effects between links are normally correlated. This phenomenon leads to new research directions in multihop-correlated shadowing. Recent research has considered such correlation effects and proposed more realistic channel models [50] [51]. In this section, however, we consider a simplified model in which the links are independent with no correlated shadowing effects.

3.4.3.1 Link Adaptation in Multihop Relay Networks

In the same WSN, communication links are usually subject to different channel conditions [52]. This is because wireless links are formed by sensor nodes that are randomly distributed in the environment, and channel fading often varies with time, geographical position, and radio frequency. Figure 3.5 illustrates an example of a multihop relay network model. In the figure, the source node and destination node are located in bad and good channel conditions, respectively, with several relay nodes distributed under different channel conditions. Information is forwarded to its destination through the chosen relays by taking the energy-saving routing protocol into consideration [53].

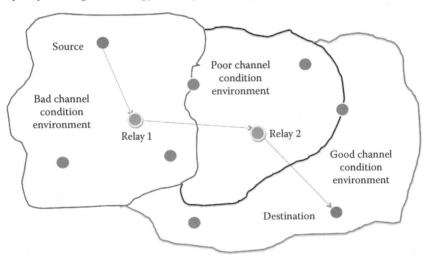

FIGURE 3.5
Example of multihop relay network.

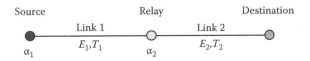

FIGURE 3.6
Two-link relay network model.

In this scenario, we are interested in analyzing the behavior of link adaptation in two-link relay networks.

A simplified model for this two-link relay network is shown in Figure 3.6. In this model, source, relay, and destination nodes form two wireless channels. The same packet is delivered through both communication links. However, transmitters in source node and relay node make independent decisions on which modulation type and power level to employ. The network is constrained by the total available energy. Although the amount of energy is fixed, the necessary transmit power allotted to each node is adaptively adjusted to achieve minimum network energy consumption.

We introduce the power allocation factor α. It is defined as the percentage of the total average power allocated to the node; hence α must satisfy $0 < \alpha < 1$. Extending Equation (3.20) and considering the scenario of discrete rate continuous power adaptation, the optimization problem can be interpreted as:

$$J(\gamma_1, \gamma_2, \ldots, \gamma_N) = \sum_{0 \leq i \leq N-1} r_i \int_{\gamma_i}^{\gamma_{i+1}} p(\gamma) d\gamma + \lambda \left[\sum_{0 \leq i \leq N-1} \int_{\gamma_i}^{\gamma_{i+1}} \frac{h(r_i)}{\gamma} p(\gamma) d\gamma - \alpha \right] \quad (3.25)$$

The optimal rate and power region boundaries can be obtained using essentially the same procedure described by Equations (3.21)–(3.23). Inserting the new parameter α to Equation (3.24), the transmit power constraint becomes:

$$\sum_{0 \leq i \leq N-1} \int_{\gamma_i}^{\gamma_{i+1}} \frac{h(r_i)}{\gamma} p(\gamma) d\gamma = \alpha \quad (3.26)$$

In the two-link relay network shown in Figure 3.6, the total energy is divided between the source node and the relay node. Denote their power allocation factors as α_1 and α_2, respectively. Both α_1 and α_2 satisfy the energy constraints in the network; that is,

$$0 < \alpha_1 < 1, 0 < \alpha_2 < 1 \quad (3.27)$$

$$\alpha_1 + \alpha_2 = 1 \quad (3.28)$$

Denote the energy consumption in link 1 and link 2 as E_1 and E_2, and the link lifetimes as T_1 and T_2, respectively. In addition, it is assumed that when the energy allocated to either node has been exhausted, the whole network is considered "dead" and reaches the limit of its lifetime. Define the network lifetime as the period of time that both nodes are alive and able to transmit data; that is, *Network Lifetime = min(Link1 lifetime, Link2 lifetime)*. The optimal network lifetime T^* is expressed as:

$$T^* = \max_{\alpha^*}\left(\min\left(T_1, T_2\right)\right) \tag{3.29}$$

where α^* is the optimal power allocation vector that contains the power allocation factors for both links when the network lifetime is maximized. Considering the energy consumption, we can also formulate the problem by calculating the minimum energy consumption in the entire network subject to optimal power allocation:

$$E = \min_{\alpha^*}\left(max(E_1, E_2)\right) \tag{3.30}$$

Thus, by solving the optimal network energy consumption problem, the optimal power allocation factors are determined.

3.4.3.2 Link Adaptation in Multiple-Link Networks

When there are several source nodes and destination nodes in the network, data transmission between nodes establishes multiple communication links. This is known as the multiple-link network. Figure 3.7 depicts a five-link example where five source nodes and five destination nodes are randomly distributed in the network. Each source node can only communicate with its own destination node and all the links operate individually. Therefore, energy consumption of each link could be considered independently. For different channel-fading conditions, the nodes consume different amounts of energy for delivering packets successfully. When a source node exhausts

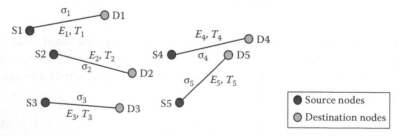

FIGURE 3.7
Example of multiple sources and destinations network model.

its battery energy, it is no longer able to transmit information and the connectivity of this particular path is lost. The network lifetime is related to the number of nodes still alive.

3.4.4 MAC Layer Adaptive Modulation and Adaptive Sleep

This section investigates a number of MAC layer [54] [55] [56] [57] protocols involving adaptive modulation and adaptive sleep.

Figure 3.8 displays the total frame duration that can be split into the actual data transmission time (labeled as Tx in the figure) and active/idle time when the transmission process has been completed and no further actions are required. Between two subsequent transmissions, the nodes can be put to sleep.

Notice that the active/idle stage is the MAC layer preallocated contention period, which ensures that the previous transmission is completed before the next one happens. During this contention period, the node can operate either in the active stage with high CPU activities or in the idle stage with low CPU activities depending on different CPU functionalities. Obviously, the active stage requires a higher power level than the idle stage, but since the radio transmission is off in the active stage, the power levels are much smaller compared to the transmission stage. In WSNs, nodes send acknowledgment (ACK) packets such as idle channel information, clear to send (CTS), and request to send (RTS) to indicate their channel status. Based on different ACKs, the node responds with corresponding actions. These actions, however, also take time to be processed, thus incurring delays and sometimes collisions. In this sense, the contention period gives extra room for the channel to deliver the previous packet and prepare the channel for the next one. After the fixed contention period, the node enters the sleep stage to wait for the next transmission.

The sleep stage, on the other hand, is the time interval between transmissions. It is also known as the "hibernation mode," during which all components in the node are not processing. Since nodes in WSNs do not need to operate at all times, they can be powered down right after completing the contention period. Energy consumption is thus minimized. Depending on how frequently the data is transmitted, the sleep time might account for

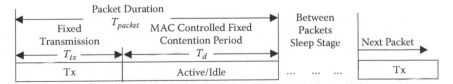

FIGURE 3.8
Packet duration in MAC layer protocol.

a substantial portion of the overall communication duration. Consequently, energy saving can be considerable.

We consider a basic MAC layer protocol in which the time of channel occupation is fixed for a given transmission. The protocol also preallocates the durations of the node transmission and the active and idle stages, giving no freedom for energy adaptation. As illustrated in Figure 3.8, every time the node detects a packet waiting to be transmitted, the protocol wakes up the node and sets the (fixed) durations for the transmission, active, and sleep stages.

Adaptive modulation techniques, on the other hand, dynamically change the data rates according to the channel conditions. This normally varies the transmission duration to be less than the preallocated transmission, hence reducing the amount of energy consumed by the transmission stage. For further optimization in energy consumption, the network can consider employing energy-aware protocol for intercommunication between sensor nodes, also known as MAC layer adaptive sleep [58]. Adaptive sleep techniques have been studied recently to improve energy efficiency in WSNs. It uses MAC layer protocols to set the duty cycle for the contention period and adaptively adjusts the sleep time according to CSI. Therefore, instead of remaining in the active or idle stages, it can stay in the sleep stage after the contention period until the next packet is ready to be transmitted.

In the following sections, we explain the operating stages for adaptive modulation (AM), adaptive modulation with idle mode (AMI), and adaptive sleep with adaptive modulation mode (ASAM). The total energy (E_t) is defined as the available battery energy per node.

The communication processes consume energy and, due to different operation stages, the total consumed energy (E_c) consists of (1) transmission energy (E_{tx}), (2) active energy (E_{active}), and (3) idle energy (E_{idle}). The sleep energy (E_{sleep}) is also considered in the process. In order to conduct a fair comparison of the energy consumption among the three algorithms, the sleep energy level and the time intervals between packets are considered the same. However, the sleep duration is dependent on the traffic intensity in the network. The time dedicated to one packet (T_{packet}) and duty cycle of the contention period (T_d) are predetermined by the MAC layer protocol.

3.4.4.1 Energy Consumption in Adaptive Modulation Mode

Combining AM techniques with the MAC layer protocol, the transmission time can be dynamically adjusted via adapting modulation schemes. As mentioned in Section 3.4.1, energy consumption is the product of the power level and transmission time. Hence, the less time for the transmission, the less energy is used during the process. AM allows the data to be transmitted at different rates according to the channel conditions. It adaptively changes the modulation constellation size, thereby varying the rate and transmis-

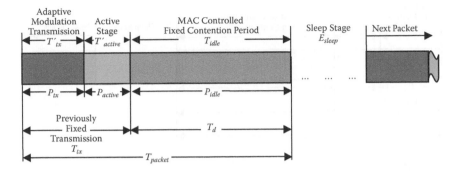

FIGURE 3.9
Energy consumption in adaptive modulation mode.

sion time. Thus, the duration for the actual transmission process does not necessarily cover the entire transmission time (T_{tx} as shown in Figure 3.8).

As illustrated in Figure 3.9, the node first switches to the active stage with a lower power level after the transmission. It then enters the idle stage, which further reduces the power level compared to the other two stages. Note that the time for the transmission stage is controlled by AM, while the node idle duration is still controlled by the MAC layer protocol. The overall energy consumption is calculated by considering the three types of energy expenditure in the network:

$$E_c^{AM} = E_{tx}' + E_{active}' + E_{idle}$$

$$= P_{tx} \times T_{tx}' + P_{active} \times T_{active}' + P_{idle} \times T_{idle}$$

$$= P_{tx} \times \frac{l}{R} + P_{active} \times (T_{packet} - \frac{l}{R} - T_d) + P_{idle} \times T_d \quad (3.31)$$

where P_{tx} is the transmit power; $T_{tx}' = \frac{l}{R}$ is the adaptive transmission time where l is the packet length and R is the AM rate; $E_{tx}' = P_{tx} \times T_{tx}'$ is the transmission energy; P_{active} is the active stage power; T_{active}' is the time the node operates in active stage, which varies in line with T_{tx}'; $E_{active}' = P_{active} \times T_{active}'$ is the active energy; P_{idle} is the idle power; T_{idle} is the fixed idle time determined by MAC layer protocol; and $E_{idle} = P_{idle} \times T_{idle}$ is the idle energy. Also note that $T_{tx}' = \frac{l}{R}$, $T_{idle} = T_d$, and $T_{tx}' + T_{active}' + T_{idle} = T_{packet}$.

3.4.4.2 Energy Consumption in Adaptive Modulation with Idle Mode

In AM, the node operates in the active mode after transmission when the CPU still has high activities, which might include executing back-stage programs and/or storing information to the memory. However, in WSNs, there is no need for the node to maintain a high volume of activities after each

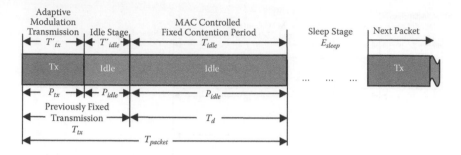

FIGURE 3.10
Energy consumption for adaptive modulation with idle mode.

transmission. Instead, it can operate in the idle mode immediately after transmission (Figure 3.10). This offers additional improvement in energy efficiency. This process is called adaptive modulation with idle mode. Its total energy consumption is determined as:

$$E_c^{AMI} = E_{tx}' + E_{idle}'$$

$$= P_{tx} \times T_{tx}' + P_{idle} \times T_{idle}' + P_{idle} \times T_{idle}$$

$$= P_{tx} \times \frac{l}{R} + P_{idle} \times \left(T_{packet} - T_d - \frac{l}{R} \right) + P_{idle} \times T_d$$

$$= P_{tx} \times \frac{l}{R} + P_{idle} \times \left(T_{packet} - \frac{l}{R} \right) \tag{3.32}$$

Similarly, P_{tx} is the transmit power; $T_{tx}' = \frac{l}{R}$ is the adaptive transmission time; l is the packet length; R is the AM rate; $E_{tx}' = P_{tx} \times T_{tx}'$ is the transmission energy; P_{idle} is the idle power; T_{idle}' is the adaptive idle time; $T_{idle} = T_d$ is the fixed node idle time; and $E_{idle}' = P_{idle} \times T_{idle}' + P_{idle} \times T_{idle}$ is the idle energy. In this case, $T_{tx}' + T_{idle}' + T_{idle} = T_{packet}$.

3.4.4.3 Energy Consumption in Adaptive Sleep with Adaptive Modulation Mode

In order to further minimize energy consumption, an ASAM algorithm is used that combines the adaptive sleep with adaptive modulation based on the MAC layer protocol. As shown in Figure 3.11, the MAC layer still allocates the same amount of idle period to ensure the previous packet was successfully delivered. However, instead of staying in the active or idle stage for a while, the node is put into the sleep stage right after the preallocated contention period T_d. In other words, the additional active or idle stage is not necessary for this algorithm. The total time for completing one packet is adaptively

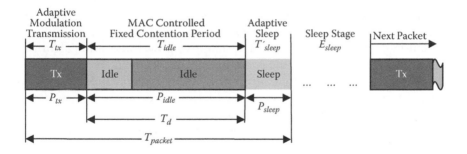

FIGURE 3.11
Energy consumption for adaptive sleep with adaptive modulation mode.

reduced. Therefore, not only the transmission time is adapted by AM, but also the sleep time is varied by AS. The energy consumption formula

$$E_c^{ASAM} = E'_{tx} + E_{idle} + E'_{sleep}$$

$$= P_{tx} \times T'_{tx} + P_{idle} \times T_{idle} + P_{sleep} \times T'_{sleep}$$

$$= P_{tx} \times \frac{l}{R} + P_{idle} \times T_d + P_{sleep} \times \left(T_{packet} - \frac{l}{R} - T_d \right) \quad (3.33)$$

Note that in this case, P_{sleep} is the sleep power, T'_{sleep} is the adaptive sleep duration, and $E'_{sleep} = P_{sleep} \times T'_{sleep}$ is the sleep energy dynamically controlled by AS protocol.

3.5 Simulation Examples and Illustrations

This section first explains the simulation steps and methodology and then presents the simulation results. The experimental parameters and assumptions are discussed. Based on the framework established in Section 3.2, we investigate the energy consumption in the network under different degrees of lognormal shadowing and Rician fading channel conditions. Different link adaptation policies are studied and used to evaluate the network lifetime. Comparisons are drawn among the AM, AMI, and ASAM schemes. The impact of different transmission parameters on node lifetime is analyzed. Furthermore, we calculate the optimal power allocation vectors for a two-link relay network and investigate its effects on the network lifetime.

3.5.1 Simulation Objective

The objective of this work is to evaluate the performance of adaptive modulation polices under energy constraints for WSNs. By considering the network constraints, various link adaptation protocols are evaluated based on energy optimization methods. The joint optimization of rate and power is performed for point-to-point communication. The adaptive power allocation algorithm is further examined for reducing energy cost in multihop networks. The second goal is to investigate the ability of the ASAM algorithms to improve energy saving. The system's operating lifetime is compared by assessing the energy efficiency using different algorithms. Finally, the performance of AM, AMI, and ASAM on commercial WSN transceivers is investigated.

3.5.1.1 Simulation Parameters

The simulation parameters are selected such that the QoS requirements for the worst-case scenario are met [59] [60]. Relevant parameters are listed in Table 3.1.

3.5.1.2 Simulation Assumptions

Trading off the simulation accuracy with complexity, the following assumptions are made:

- Both transmitter and receiver have perfect knowledge of CSI.
- The transmission parameters remain constant over one frame period.
- Channel conditions stay the same during transmission.
- No errors are introduced by the feedback channel, and the delay is negligible.
- The transmission process is reliable and all packets arrive independently.
- Perfect routing nodes are selected and all nodes are aware of their position to others.

3.5.2 Energy Optimization

The energy efficiency in the network is highly dependent on the transmission parameters and channel conditions. The focus of the simulation is to evaluate different energy optimization algorithms by adjusting the network parameters. For a given BER constraint, the optimal energy in the network is obtained by jointly considering the data rate and the transmit power. Since nodes in a network can be distributed among different channel condition environments, the channel conditions are modeled by considering different fading effects for each link.

Optimizing the data rate and the transmit power jointly contributes to energy cost reduction in transmission under fading channel conditions. Given the set of MQAM constellations and BER constraints listed in Table 3.1, the switching levels for the discrete rate adaptation can be calculated using Equation (3.11).

TABLE 3.1

Simulation Parameters

Parameter	Description	Value
K	Rician K-factor	{0, 5, 10, 15, 20} dB
σ	Lognormal shadowing variance	{0, 2, 4, 6, 8} dB
$\bar{\gamma}$	Average SNR	10–36 dB
P_b	BER requirement for QoS	10^{-4}
M	MQAM modulation constellation size	{2, 4, 16, 64, 256}
B	Channel bandwidth	200 kHz
n	Path-loss exponent	3
E_t	Total battery capacity	1200 mAhr
V	Operating voltage	3.6 V
I_{Tx}	Transmit current	120 mA
I_{active}	Active current	100 mA
I_{idle}	Idle current	1 mA
I_{sleep}	Sleep current	0.1 mA
T_d	Duty cycle for contention period	75%
T_{packet}	Packet duration	1 unit of time

The transmitter is designed to automatically switch to the next level modulation scheme once the average SNR exceeds the threshold. However, within the same modulation scheme, the transmission rate stays constant. Using Equations (3.22) and (3.23), the optimal transmission rates are obtained. Packets are then delivered at the optimal rate to decrease the transmission time.

3.5.2.1 Energy Consumption in AM, AMI, and ASAM

This section investigates the energy consumption in AM, AMI, and ASAM. Recalling the energy consumption formulas, the node lifetime of AM, AMI, and ASAM is calculated based on the different operating stages required by each algorithm. We analyze the source node lifetime for successfully delivering a certain number of packets using the adaptive discrete rate continuous power policy.

The node lifetime achieved by the three algorithms under different degrees of channel-fading conditions are shown in Figure 3.12 for lognormal shadowing and Figure 3.13 for Rician fading, respectively. As can be seen from the figures, the ASAM algorithm consistently outperforms the AM technique. Nodes can operate up to 221 days using ASAM, while the longest node lifetime using AM is only 42 days. This translates into an approximately 80% improvement. In addition, AMI also slightly improves the node lifetime compared to AM. At the high SNR region, AMI extends the node lifetime by approximately 10 days.

In AM, the node is first active after the transmission stage, then switches to the idle stage, and finally enters the sleep stage. In AMI, the active stage is

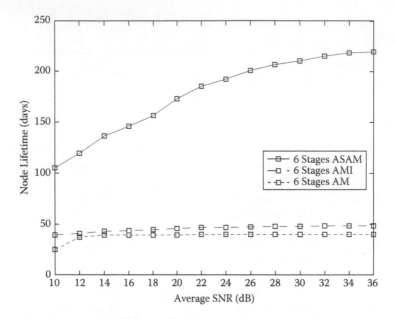

FIGURE 3.12
Node lifetime simulation: comparison of six stages ASAM, AMI, and AM for lognormal shadowing, $\sigma = 2$ dB.

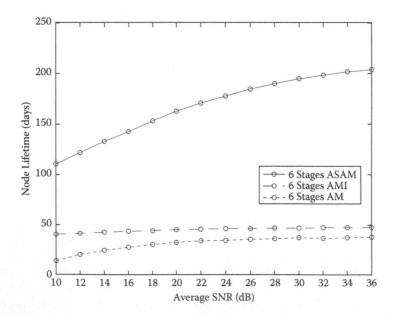

FIGURE 3.13
Node lifetime simulation: comparison of six stages ASAM, AMI, and AM for Rician fading, $K = 5$ dB.

replaced by the idle stage since the processor only requires minimum functionality. Therefore, this portion of the energy is reduced from active energy to idle energy. However, since the power level at the sleep stage is the least, it would be desirable to have the node stay in this stage for the maximum period of time. The ASAM algorithm provides an optimal adjustment of node sleep duration according to the CSI, and therefore further enhances the energy efficiency. The main difference between ASAM and AM is that the former enters the sleep stage right after the contention period that is preallocated by the MAC layer protocol, while the latter stays active for a certain period of time before entering the contention period. The significant difference between the sleep power and active power stages can considerably affect node lifetime, as verified in the figures.

Furthermore, the results also indicate that there are a number of factors that can affect the node lifetime. The factors investigated in this section include: channel fading, average SNR, traffic intensity, and modulation stages. They are discussed in detail in the following sections.

3.5.2.2 Channel Fading and Average SNR

Different channel-fading conditions exert different influences on per-node energy consumption [56], as examined in this section.

The transmission protocol allows the transmitter to choose from among six modulation stages: no transmission, BPSK, 4QAM, 16QAM, 64QAM, and 256QAM. Here we consider a slowly varying flat-fading channel. The channel changes much slower than the symbol data rate, so that it remains approximately constant over each transmission period. Lognormal shadowing and Rician distribution are used to model the channel fading effects, respectively. For lognormal distribution, the standard deviation σ is chosen from among $\{0, 2, 4, 6, 8\}$; for Rician distribution, the K-factor K is taken from $\{0, 5, 10, 15, 20\}$.

Figure 3.14 and Figure 3.15 display the node lifetime as a function of σ and K, for lognormal shadowing and Rician fading effects, respectively. Figure 3.14 clearly shows that for each algorithm, under a given average SNR, stronger lognormal shadowing (that is, higher σ) yields a shorter node lifetime. In other words, the energy cost is reduced when the network operates in a good channel condition. The behavior is similar for the Rician fading case (see Figure 3.15). The node can operate longer when the fading in the wireless channel is minor (that is, higher K). The Rician K-factor determines the strength of the line-of-sight (LOS) for the point-to-point communication. Hence, a smaller K-factor gives deeper fading in the channel and shorter lifetime as a result. Moreover, the figures also demonstrate that the node lifetime is heavily dependent on the average SNRs in the network. For the same channel-fading condition, increasing the average SNR extends the node lifetime. This trend applies to both lognormal shadowing and Rician fading

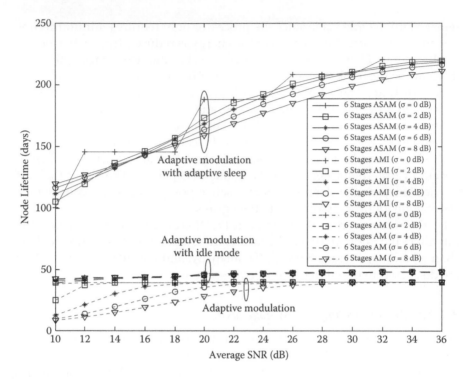

FIGURE 3.14
Node lifetime simulation: comparison of six stages ASAM, AMI, and AM under various log-normal shadowing channel conditions.

channel conditions. For strong fading cases, a poor SNR can substantially degrade node lifetime performance.

Although the figures indicate that for all algorithms the sensor nodes can operate longer with smaller channel fading and larger SNRs, the channel-fading conditions and average SNRs show very distinct impacts on energy consumption for different algorithms. For AM, the channel condition has a higher impact on the node lifetime at low SNRs than at high SNRs. As the average SNR increases, the lifetime gap between different fading conditions becomes smaller. For instance, in Figure 3.14, the node lifetime curves roughly overlap with each other for average SNR greater than 30 dB. We can also conclude that the average SNR values become the dominant factor for the energy consumption when the node operates under deep channel fading. However, unlike in AM where good channel conditions considerably improve node lifetime, the channel-fading conditions and average SNR values only affect the energy consumption slightly in AMI.

Such behaviors are due to the fact that in AM, after the packet has been successfully transmitted, the channel still stays active for a certain period of time. For nodes operating in a poor channel condition, there are higher

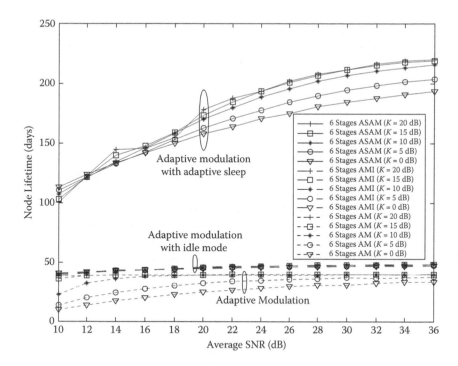

FIGURE 3.15
Node lifetime simulation: comparison of six stages ASAM, AMI, and AM under various Rician fading channel conditions.

chances of resending the packet. This resending incurs more active power consumption of the node. On the other hand, AMI puts the node into the idle stage immediately after transmission. Even under poor channel conditions where a lot of retransmissions occur, the node only spends the idle power after the retransmission, which is normally much smaller than active power. The difference in energy consumption between various levels of degrading channel fading is thus not notable in AMI.

The channel-fading conditions and average SNRs also have significant impacts on the node lifetime in ASAM. Compared to AM, the performance of the ASAM algorithm is more notably influenced by the average SNR. The difference in node lifetime is approximately 120 days between the lowest and the highest SNRs for ASAM, while for AM, the difference is less than 30 days. In addition, the node lifetime is more susceptible to channel condition variations in low and high SNR regions for AM and ASAM, respectively.

Interestingly, it can also be seen from Figure 3.14 and Figure 3.15 that for AMI and ASAM, in the low average SNR region, the node lifetime with larger σ turns out to be longer than smaller σ cases. Recall that σ represents the standard deviation of the instantaneous SNRs. Under low average SNR, a larger σ indicates both a higher chance of retransmission and a higher chance of transmission with a larger data rate. Retransmission incurs additional energy consumption, while a higher data rate reduces the energy cost by using less transmission time. For AMI and ASAM, the retransmission cost is essentially very small; hence the energy saving gained from higher-rate transmission outweighs the downside—that is, the larger σ, the longer node lifetime, as shown in the plots. For AM cases, on the other hand, a smaller σ consistently yields better performance relative to a larger lifetime. Such a phenomenon happens because the retransmission cost for AM is considerably higher than for AMI/ASAM due to the active power level used when retransmission occurs. The disadvantage of having larger σ outweighs the advantage in such cases.

Moreover, under a high average SNR, the node is already transmitting data with a high rate. Increasing σ, therefore, might contribute a small degree to a better transmission speed but would increase the likelihood of having low instantaneous SNR, that is, a low data transmission rate and more energy consumption. As reflected in the figures, a smaller sigma always gives a longer node lifetime in the high average SNR region.

3.5.2.3 Traffic Intensity

Traffic intensity is a measure of the average occupancy of wireless resources during a period of time. Depending on the traffic intensity, the energy expenditure in the network can differ considerably. In this section, we further investigate energy consumption using AM, AMI, and ASAM with different profiles of network traffic intensity, namely, 1%, 10%, and 100%. The network traffic is modeled using Poisson random distribution. A larger intensity indicates more packets that need to be transmitted in the same period of communication time.

Figure 3.16 and Figure 3.17 depict the node lifetime in AM under different traffic intensity profiles for lognormal shadowing and Rician fading, respectively. It is evident from the figures that for low and moderate average SNRs, low traffic communications yield higher node lifetime. For a given traffic intensity, the node lifetime curves for different fading levels tend to converge as the average SNR increases. The converged node lifetime with 1% traffic intensity can be roughly three to five times as long as the 100% intensity case, depending on the fading type. Also, the convergence speed appears to be dependent on the traffic intensity. The heavier the traffic, the larger the average SNR that is required for the lifetime curves to reach convergence.

Similarly, the impact of traffic intensity on energy consumption is studied for AMI. Figure 3.18 and Figure 3.19 illustrate how the AMI algorithm performs with respect to different traffic loads. As shown in the figures, a heavy traffic load in the channel reduces the sensor node lifetime in AMI, similar to the AM case. Nodes can operate up to approximately 49 days under 1% traffic intensity, while the time is reduced to 46 days and 38 days under 10% and 100% traffic intensity, respectively.

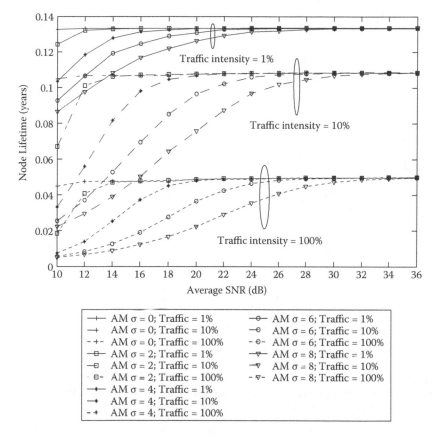

FIGURE 3.16

Impact of traffic intensity on node lifetime using AM under lognormal shadowing. Traffic in the channel is 1%, 10%, and 100% of the load.

The results of ASAM algorithm are given in Figure 3.20 and Figure 3.21. The node lifetime can reach up to 439 days under a low traffic profile, which is almost 10 times longer than the node lifetime in AM and AMI. When the network traffic load gets heavier to 10% intensity, the node lifetime is reduced by nearly half to 220 days. Even under the full load (that is, 100% traffic intensity), the node can still operate over 50 days, which is higher than AM and AMI.

For all algorithms, it can be concluded that the node lifetime performance is highly dependent on the number of transmissions in the network. This result is expected because when the network has a heavier traffic load, it implies that more packets are in the queue waiting to be delivered. To complete the transmissions of all these packets, the node has to drain more energy from its battery, resulting in a shorter lifetime. In the remainder of the chapter, the traffic intensity is assumed to be equal to 100%.

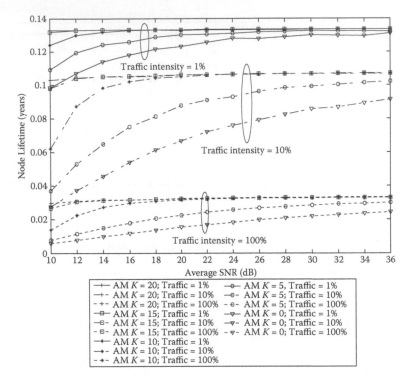

FIGURE 3.17
Impact of traffic intensity on node lifetime using AM under Rician fading. Traffic in the channel is 1%, 10%, and 100% of the load.

3.5.2.4 Modulation Stages

The number of stages used in AM varies from application to application and is also subject to the complexity of practical implementation. In this experiment, we compare energy efficiency using AM, AMI, and ASAM with different numbers of modulation stages. Four scenarios are compared: (1) three stages (no transmission, BPSK, QPSK), (2) four stages (no transmission, BPSK, QPSK, 16QAM), (3) five stages (no transmission, BPSK, QPSK, 16QAM, 64QAM), and (4) six stages (no transmission, BPSK, QPSK, 16QAM, 64QAM, 256QAM).

The node lifetime for the four scenarios is illustrated in Figure 3.22 and Figure 3.23. As can be seen, with more modulation stages built into the transmitter, the sensor node can operate for a longer period of time. This behavior is due to the fact that modulation levels restrict the transmission rate. Transmitters that can only choose from low-level modulation schemes have to use a smaller data rate even under very good channel conditions, hence increasing the transmission duration. On the other hand, when the transmitter is equipped with the possibility to choose from higher-level modulation schemes, transmissions benefit from a higher achievable data rate under

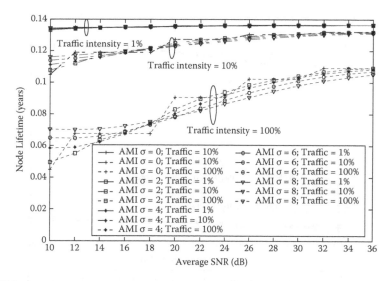

FIGURE 3.18
Impact of traffic intensity on node lifetime using AMI under lognormal shadowing. Traffic in the channel is 1%, 10%, and 100% of the load.

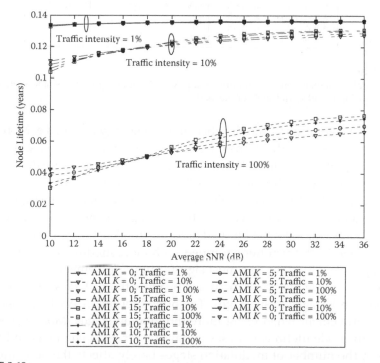

FIGURE 3.19
Impact of traffic intensity on node lifetime using AMI under Rician fading. Traffic in the channel is 1%, 10%, and 100% of the load.

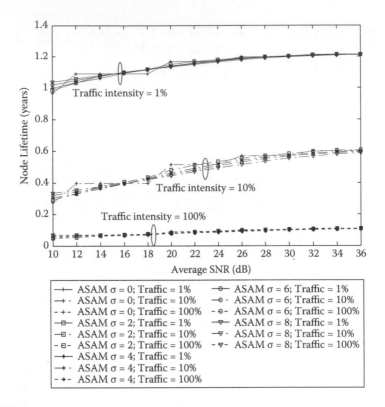

FIGURE 3.20
Impact of traffic intensity on node lifetime using ASAM under lognormal shadowing. Traffic in the channel is 1%, 10%, and 100% of the load.

good channel conditions and thus become less energy expensive. Moreover, it can also be observed that transmitters with more modulation stages exhibit an advantage after certain SNR thresholds. These thresholds are determined by the discrete rate continuous power link adaptation policy.

Comparing AM, AMI, and ASAM, it is clearly indicated in the figures that the number of modulation stages has the most significant impact on the node lifetime in ASAM, followed by AMI, and has the least important effects in AM. In general, a transmitter equipped with more modulation stages has a longer operating lifetime. The most significant improvement is achieved by ASAM, as the node lifetime is increased by more than 70 days when increasing from three stages to six stages. The node lifetime in AMI also shows a slight improvement when using more modulation stages. More stages can increase the node lifetime up to seven days under good channel conditions. However, the number of modulation stages barely affects the node lifetime when AM is used, with less than one day improvement gained under the same condition.

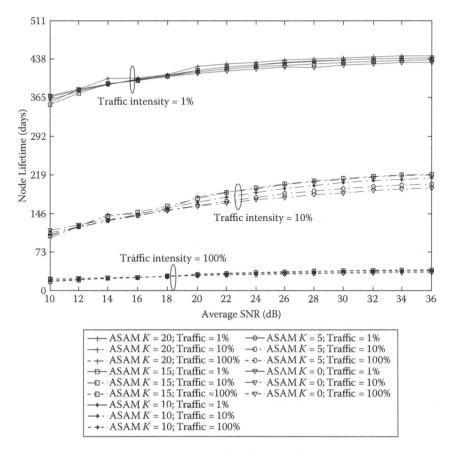

FIGURE 3.21
Impact of traffic intensity on node lifetime using ASAM under Rician fading. Traffic in the channel is 1%, 10%, and 100% of the load.

Such behaviors can be interpreted as follows. The major benefit of using more modulation stages lies in the increased capability of taking advantage of the good channel conditions to reduce transmission time. During the saved amount of time, the node can switch from the most power-hungry transmission stage to some other stage, as determined by the specific algorithm. Between ASAM and AMI, the former adapts the duration of the sleep stage; while in the latter, the node still needs to operate in the idle stage, which consumes an extra portion of idle energy. For AM, however, the node operates in the active stage after the transmission, in which the power level, although lower than transmission power, is still much higher than the sleep and idle power levels. As a result, even if the transmitter is able to choose modulation levels with higher data rates in AM, the energy saving is relatively insignificant compared to ASAM or AMI.

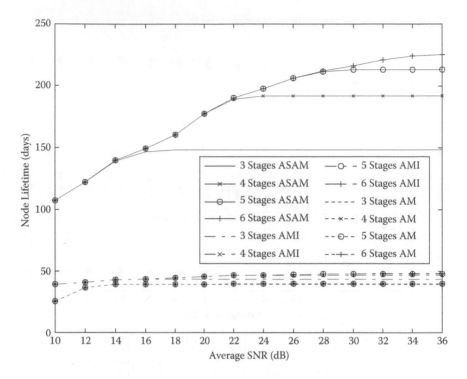

FIGURE 3.22
Impact of modulation stages on node lifetime using ASAM, AMI, and AM under lognormal shadowing when σ = 2 dB.

3.5.2.5 Discussion of Energy Consumption

The simulation results indicate that ASAM outperforms AM and AMI by reducing more energy consumption. In general, the impacts of transmission parameters on the node lifetime are found to be similar for AM, AMI, and ASAM. For all the algorithms, lognormal shadowing and Rician fading conditions show similar impacts on node lifetime. The node can operate longer when the wireless link is located in a good channel condition with minor fading and high average SNRs. A lower traffic load for communication and more modulation stages in the transmitter also contribute to lifetime improvement.

However, there are still a few differences, as suggested by the results. Comparing the energy consumption under different degrees of fading effects, it can be found that the node lifetime can substantially benefit from good channel conditions in AM and ASAM. In AM, less fading in the channel can extend the node lifetime by approximately 30 days, compared to the case of deep fading, while for ASAM, the maximum increment of the node lifetime can reach up to 20 days. For AMI, on the other hand, the impact

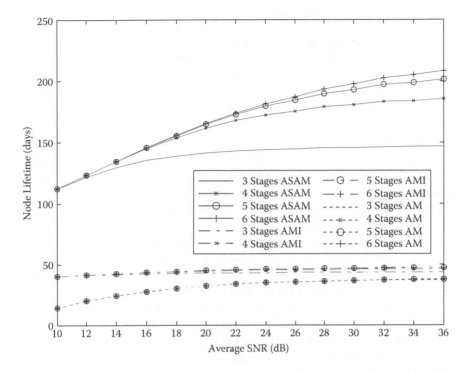

FIGURE 3.23
Impact of modulation stages on node lifetime using ASAM, AMI, and AM under Rician fading when K = 5 dB.

of channel conditions on the node lifetime is insignificant. In addition, for AM, variations in channel-fading conditions affect the node lifetime considerably under low SNRs. For ASAM, however, the effect of channel fading is basically consistent when varying the average SNR; hence both channel fading and average SNR play important roles in energy consumption in the network.

Furthermore, the number of modulation stages is found to have considerable impact on the node lifetime for ASAM. The node lifetime can be improved by up to 70 days when switching from three stages of modulation to six stages of modulation. For AMI, using more modulation stages in the transmitter helps to improve node lifetime by seven days. However, the improvement attained in AM is minor. Using more modulation stages slightly increases the operating time by one day under good channel conditions; the improvement is even smaller under poor channel conditions. Moreover, to build a large number of modulation schemes in one transmitter increases the hardware implementation complexity. When designing AM protocols, a tradeoff has to be considered between the energy consumption reduction obtained by using more modulation stages and the practical implementation complexity and cost.

3.5.3 Power Control Adaptation Policies

In this section, we investigate adaptive power control policies in single-hop networks. The six modulation stages are available to the link adaptation and target BER is set to 10^{-4}, and the average SNR is varied from 10 dB to 36 dB.

Since the power level increases rapidly only when the transmitter is switching to the next modulation stage (Section 3.3.5.3), the transmit power used in the communication processes does not necessarily remain as a constant at the highest level. A power control factor is introduced to adjust the value of transmit power within the same level of modulation. Based on Equation (3.6), the optimal values of the power adaptation control factor are derived as shown in Figure 3.4.

In this analysis, power and rate are jointly adapted according to channel conditions. Link adaptation with an optimal discrete rate and an optimal power control is considered in order to achieve maximal spectral efficiency under a given average BER constraint. The node lifetime is compared between the following six cases: (1) adaptive sleep combined with adaptive modulation under power control (ASAM-PC), (2) adaptive sleep combined with adaptive modulation, no power control (ASAM), (3) adaptive modulation with idle mode under power control (AMI-PC), (4) adaptive modulation with idle mode, no power control (AMI), (5) adaptive modulation under power control (AM-PC), and (6) adaptive modulation, no power control (AM). Different fading levels are still considered.

Figure 3.24 and Figure 3.25 show superior performance of ASAM, which is due to the fact that it has a very good adaptation capability given the algorithm's ability to dynamically adjust the operating durations for both the transmission stage and the sleep stage. However, power control results in insignificant improvements in AMI and AM cases because nodes only adapt the duration of the transmission stage without taking the active and idle stages into consideration, respectively. Even though the power control algorithm is able to lower the power level of the transmission stage, the portion of active energy and idle energy still dominates the total energy consumption in the network. Therefore, in most cases, the effect of power control would be more noticeable for ASAM than for AM and AMI.

3.5.4 Two-Link Relay Network Adaptation

When there are multiple nodes in the network, the communication between nodes can be categorized into two types: (1) multihop relay networks that transmit the source data to the destination through relay nodes, and (2) multiple-link networks where multiple sources and destinations exist in the network and each source transmits data to its destination independently. Here we investigate the energy consumption using link adaptation in two-link relay networks.

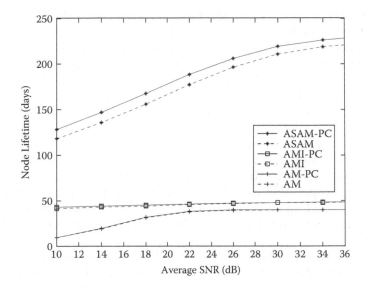

FIGURE 3.24
Power control impacts on node lifetime using six stages ASAM and AM under the lognormal shadowing when σ = 6 dB.

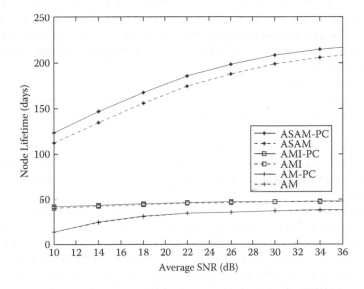

FIGURE 3.25
Power control impacts on node lifetime using six stages ASAM and AM under Rician fading when K = 5 dB.

Recall the relay network link adaptation control explained in Section 3.4.3.1. The proposed adaptive power allocation policy is employed here. Network lifetime depends on the energy consumption of all wireless links along the communication path. Although the total available energy in the network is finite, the power allocated to each node can be adjusted to achieve optimal transmission quality with minimal energy consumption.

This simulation considers the simplest relay network with one relay node so that information will always be routed through the relay node to the destination. Hence two links are created in the network: source to relay and relay to destination. Due to the wireless multipath and scattering, the two wireless links can have different fading effects. The power allocation policy, therefore, is responsible for allotting the optimal power according to the instantaneous fading information of the links. The optimal power allocation algorithm is compared to the case in which the total energy is evenly divided between the links.

The optimal power allocation factors of the two links are calculated and shown in Table 3.2 and Table 3.3 for cases of lognormal shadowing and Rician fading, respectively. These values also corroborate the previous observations that ASAM benefits more from power allocation than AM. In general, the allocated power is biased to the link with inferior channel conditions. This is due to the fact that the energy resources must be allocated in such a way that operations of all links in the network are maintained simultaneously for the maximal period of time. Normally, the node with the worst channel conditions has low energy efficiency and fails first. Recall Equation (3.29). Since a network is considered functional only if both links are alive, the minimum of the lifetime of the two links needs to be maximized. Such a phenomenon will naturally grant more energy to the link located in poorer channel conditions with higher energy expenditure; by extending its lifetime, the lifetime of the entire network is extended.

It can be seen from Figure 3.26 and Figure 3.27 that for the two-link relay networks, improved node lifetime is obtained for all the algorithms with the optimal power allocation. The improvement is more notable for the combination of the power allocation algorithm and ASAM. As the node lifetime using ASAM is more subject to channel conditions than that using AM, wisely allocating the energy to each link according to its channel fading is essential for enhancing the energy efficiency of the entire network in ASAM.

Moreover, the results also indicate that a more considerable increment can be achieved under more distinct fading conditions between the two links. Also, power allocation with the AM scheme is shown to perform better at low SNR regions than at high SNR regions; with ASAM, the power allocation algorithm is found to improve the network lifetime more at high SNRs. As mentioned in Section 3.5.2.2, at high SNR regions, the node lifetime under different degrees of fading tends to converge in AM; hence variations in channel-fading conditions only exert minor influence on energy

TABLE 3.2

Optimal Power Allocation Factors for Two Links: $\sigma1 = 2$ dB, $\sigma2 = \{0, 4, 6, 8\}$ dB

Optimal Allocation Factor	Method	SNR							
		10 dB	14 dB	18 dB	22 dB	26 dB	30 dB	34 dB	38 dB
Two Links Channel Condition: $\sigma1 = 2$ dB, $\sigma2 = 0$ dB									
$\alpha1$	ASAM	0.56	0.56	0.55	0.54	0.53	0.51	0.51	0.5
$\alpha2$		0.44	0.44	0.45	0.46	0.47	0.49	0.49	0.5
$\alpha1$	AMI	0.51	0.51	0.51	0.5	0.5	0.5	0.5	0.5
$\alpha2$		0.49	0.49	0.49	0.5	0.5	0.5	0.5	0.5
$\alpha1$	AM	0.56	0.55	0.54	0.53	0.53	0.51	0.5	0.5
$\alpha2$		0.44	0.45	0.46	0.47	0.47	0.49	0.5	0.5
Two Links Channel Condition: $\sigma1 = 2$ dB, $\sigma2 = 4$ dB									
$\alpha1$	ASAM	0.44	0.44	0.45	0.46	0.47	0.49	0.49	0.51
$\alpha2$		0.56	0.56	0.55	0.54	0.53	0.51	0.51	0.49
$\alpha1$	AMI	0.49	0.49	0.5	0.5	0.5	0.5	0.5	0.5
$\alpha2$		0.51	0.51	0.5	0.5	0.5	0.5	0.5	0.5
$\alpha1$	AM	0.33	0.43	0.49	0.49	0.5	0.5	0.5	0.5
$\alpha2$		0.57	0.55	0.51	0.51	0.5	0.5	0.5	0.5
Two Links Channel Condition: $\sigma1 = 2$ dB, $\sigma2 = 6$ dB									
$\alpha1$	ASAM	0.45	0.41	0.39	0.38	0.35	0.32	0.3	0.29
$\alpha2$		0.55	0.59	0.61	0.62	0.65	0.68	0.7	0.71
$\alpha1$	AMI	0.49	0.49	0.49	0.5	0.5	0.5	0.5	0.5
$\alpha2$		0.51	0.51	0.51	0.5	0.5	0.5	0.5	0.5
$\alpha1$	AM	0.32	0.33	0.44	0.49	0.5	0.5	0.5	0.5
$\alpha2$		0.68	0.67	0.56	0.51	0.5	0.5	0.5	0.5
Two Links Channel Condition: $\sigma1 = 2$ dB, $\sigma2 = 8$ dB									
$\alpha1$	ASAM	0.43	0.4	0.36	0.31	0.27	0.23	0.21	0.19
$\alpha2$		0.57	0.6	0.64	0.69	0.73	0.77	0.79	0.81
$\alpha1$	AMI	0.48	0.49	0.49	0.49	0.5	0.5	0.5	0.5
$\alpha2$		0.52	0.51	0.51	0.51	0.5	0.5	0.5	0.5
$\alpha1$	AM	0.27	0.28	0.38	0.45	0.48	0.5	0.5	0.5
$\alpha2$		0.73	0.72	0.62	0.55	0.52	0.5	0.5	0.5

consumption. In the low SNR cases, however, node lifetime is much more sensitive to fading conditions; as a result, energy resources need to be more carefully allocated to improve energy efficiency. Conversely, for ASAM, the channel fading has more significant impacts on node lifetime under high average SNRs. Therefore, adaptive power allocation is more demanding as SNR increases in this scenario.

TABLE 3.3

Optimal Power Allocation Factors for Two Links: K1 = 10 dB, K2 = {0, 5, 15, 20} dB

Optimal Allocation Factor	Method	SNR							
		10 dB	14 dB	18 dB	22 dB	26 dB	30 dB	34 dB	38 dB
Two Links Channel Condition: K1 = 10 dB, K2 = 0 dB									
α1	ASAM	0.43	0.4	0.36	0.31	0.27	0.23	0.21	0.19
α2		0.57	0.6	0.64	0.69	0.73	0.77	0.79	0.81
α1	AMI	0.47	0.48	0.48	0.48	0.49	0.49	0.5	0.5
α2		0.53	0.52	0.52	0.52	0.51	0.51	0.5	0.5
α1	AM	0.31	0.32	0.37	0.41	0.42	0.44	0.45	0.46
α2		0.69	0.68	0.63	0.59	0.58	0.56	0.55	0.54
Two Links Channel Condition: K1 = 10 dB, K2 = 5 dB									
α1	ASAM	0.47	0.46	0.44	0.42	0.39	0.37	0.36	0.35
α2		0.53	0.54	0.56	0.58	0.61	0.63	0.64	0.65
α1	AMI	0.48	0.48	0.49	0.49	0.49	0.5	0.5	0.5
α2		0.52	0.52	0.51	0.51	0.51	0.5	0.5	0.5
α1	AM	0.37	0.4	0.44	0.47	0.47	0.48	0.49	0.49
α2		0.63	0.6	0.56	0.53	0.53	0.52	0.51	0.51
Two Links Channel Condition: K1 = 10 dB, K2 = 15 dB									
α1	ASAM	0.52	0.55	0.56	0.58	0.59	0.6	0.62	0.64
α2		0.48	0.45	0.44	0.42	0.41	0.4	0.38	0.36
α1	AMI	0.52	0.52	0.52	0.51	0.51	0.51	0.5	0.5
α2		0.48	0.48	0.48	0.49	0.49	0.49	0.5	0.5
α1	AM	0.61	0.54	0.53	0.51	0.51	0.51	0.5	0.5
α2		0.39	0.46	0.47	0.49	0.49	0.49	0.5	0.5
Two Links Channel Condition: K1 = 10 dB, K2 = 20 dB									
α1	ASAM	0.53	0.56	0.57	0.59	0.61	0.64	0.65	0.68
α2		0.47	0.44	0.43	0.41	0.39	0.36	0.35	0.32
α1	AMI	0.53	0.53	0.52	0.52	0.52	0.51	0.5	0.5
α2		0.47	0.47	0.48	0.48	0.48	0.49	0.5	0.5
α1	AM	0.67	0.63	0.59	0.54	0.52	0.51	0.51	0.51
α2		0.33	0.37	0.41	0.46	0.48	0.49	0.49	0.49

3.5.5 Performance of Commercial WSN Nodes

Some published works have examined the performance of IEEE 802.15.4 transceivers and measured the current values drained from the power source for different operating modes [61] [62] [63]. Based on these empirical characteristics of battery consumption, node lifetime is calculated as a function of the operating current of the sensor nodes. Three commercial wireless transceivers are evaluated here, namely, CC2430, CC2520, and MC1322. For the

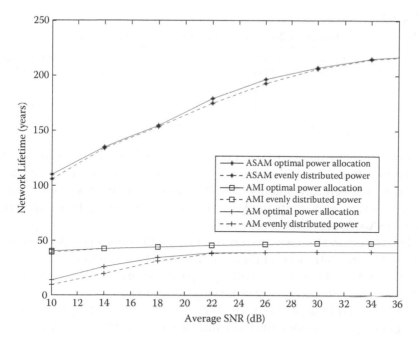

FIGURE 3.26
Power allocation under lognormal shadowing. Two-link channel conditions: σ1 = 2 dB, σ2 = 6 dB.

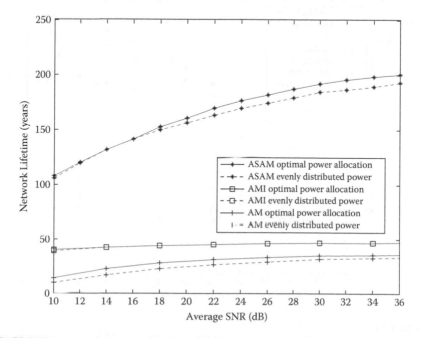

FIGURE 3.27
Power allocation under Rician fading. Two-link channel conditions: K1 – 10 dB, K2 = 0 dB.

TABLE 3.4

Current Consumption of CC2480, CC2520, and MC1322 Transceivers

Transceivers	Operating Voltage (V)	Transmission Stage (mA)	Active Stage (mA)	Idle Stage (mA)	Sleep Stage (uA)
CC2430	2.0–3.6	27	12.3	0.19	0.3
CC2520	1.8–3.6	25.8	10	0.175	0.03
MC1322	2.0–3.6	32	15	0.9	0.3

purpose of energy saving, the commercial IEEE 802.15.4 transceivers are able to switch between operating modes with different current consumptions. Here we are specifically interested in the current consumption of the transmission, active, and sleep modes. The functionalities of these operating modes are explained in Section 3.4.1.

Table 3.4 displays the current consumption for the three operating modes as provided by the datasheets of the transceivers, which is based on the experimental results. Total available battery capacity for each node is assumed to be 1200 mAhr, and packets are sent every 10 seconds.

Figure 3.28 and Figure 3.29 show that ASAM consistently yields the longest operating time among the three link adaptation techniques. Comparing AMI and AM, although the former slightly improves the node lifetime, the

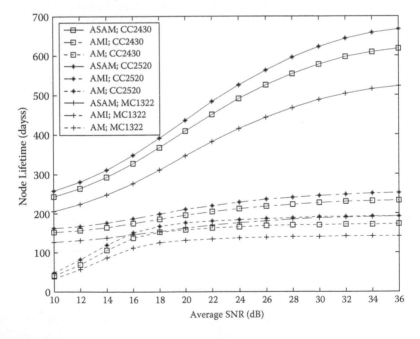

FIGURE 3.28

Comparison of the node lifetime of commercial IEEE 802.15.4–compliant transceivers using six stages ASAM, AMI, and AM under lognormal shadowing when $\sigma = 4$ dB.

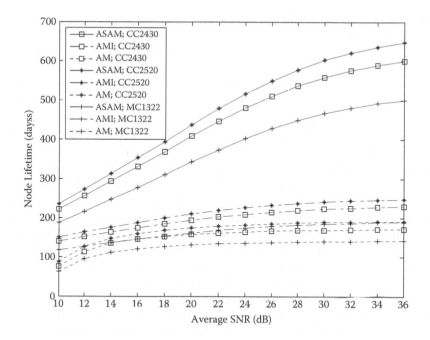

FIGURE 3.29
Comparison of the node lifetime of commercial IEEE 802.15.4–compliant transceivers using six stages ASAM, AMI, and AM under Rician fading when K = 10 dB.

performances of AMI and AM are in general very close. The trends of the node lifetime performance using commercial transceivers are very similar to the previously simulated transmitter model. This indicates that the proposed ASAM algorithm is a promising technique for improving node operating lifetime for commercial IEEE 802.15.4–compliant transceivers. In addition, good channel conditions also improve node lifetime since smaller fading effects rarely cause retransmissions.

Furthermore, since the transmitters have different current consumptions for the transmission, active, idle, and sleep stages, these values influence the power levels and therefore affect the total energy consumption of the nodes. Thus it is also worthwhile to analyze the node lifetime variations that are caused by different current consumption. It is observed that all the commercial transmitters have much lower current consumption values than the simulated transmitter model (see Table 3.1 for the parameters' values). As listed in Table 3.4, CC2520 has the lowest overall current consumption for all operating modes, while MC1322 has the highest consumption.

The node lifetime of all commercial transceivers is overall longer than the simulated transmitter model, given the smaller current consumptions in these commercial products. The CC2520 transceiver has the longest node lifetime since it operates at the lowest overall current consumption. Under good channel conditions, it can operate for over 625 days using the ASAM algorithm, while it can still last nearly 200 days and 250 days using the AM and AMI, respectively. CC2430 has moderate node lifetime, relatively close to the node lifetime of CC2520. For the same channel conditions, CC2430 operates for up to 615 days using the ASAM algorithm, and almost 245 and 181 days using the AMI and AM techniques, respectively. The MC1322 transceiver gives the shortest node lifetime: 510 days, 198 days, and 146 days, using the ASAM, AMI, and AM, respectively. This is expected because both CC2520 and CC2430 transceivers consume low current in the idle stage. Compared to MC1322, the idle mode current consumption can be reduced to nearly one fifth of the current in CC2520 and CC2430. Although CC2520 and CC2430 have very close idle mode currents, their current consumption in the transmission and active modes is still different. The higher current consumption in CC2430 leads to a shorter node lifetime relative to CC2520.

Among current values of all operating modes, the consumption during the idle mode is the most significant factor determining node lifetime. However, the current drained for transmission and active modes also shows a small impact. Moreover, current consumption during the sleep stage also has considerable impact on the node lifetime, particularly for the ASAM algorithm. Since it adaptively adjusts the duration of the sleep stage, reduced sleep current consumption can further improve the node lifetime in ASAM.

3.6 Conclusion

This chapter investigated the PHY layer characteristics and implementation issues of AM and ASAM for energy-constrained wireless sensor networks. We explained in detail the main components of the link adaptation feedback system model and discussed several AM-based techniques. The energy consumption of the adaptive protocols in various networks was evaluated. An ASAM algorithm was developed and incorporated into the link adaptation policies. The proposed algorithm was compared with the AM scheme. In addition, the optimal power allocation values for multihop relay networks were computed and the maximum network lifetime was compared between several link adaptive protocols.

The simulation results indicate that discrete rate continuous power adaptation protocol can effectively control energy consumption in the energy-constrained network where all nodes share a finite amount of energy available to the system. It is shown that AM is a promising technique increasing the data rate, while AS notably enhances power efficiency. The combination of the two can reduce energy consumption in the network, thereby improving the system's operating lifetime. The performance of point-to-point communication and multihop networks was investigated. The power allocation scheme was compared to other link adaptation protocols for multihop relay networks, and promising results have been obtained.

Furthermore, it has been shown that various transmission parameters had different impacts on energy consumption for AM, AMI, and ASAM. Channel-fading conditions have a large impact on energy consumption for AM, especially at low SNR regions. However, for ASAM, both channel-fading conditions and average SNR values show significant impact on the node lifetime. Additionally, a high modulation stage improves node lifetime more in ASAM than in AMI and AM. The two-link multihop relay network model was investigated and the relation between power allocation and energy consumption has been explored. By employing the optimal power allocation algorithm, network operating lifetime could be substantially improved in multihop relay networks.

References

1. S. Phoha et al., *Sensor network operations*. Hoboken, N.J., Wiley; IEEE Press, 2006.
2. S.G. Cui et al., "Energy-constrained modulation optimization," *IEEE Transactions on Wireless Communications* 4, September 2005, pp. 2349–2360.
3. T. He et al., "Achieving real-time target tracking using wireless sensor networks," *Proceedings of the 12th IEEE Real-Time and Embedded Technology and Applications Symposium*, 2006, pp. 37–48.
4. H. Karl et al., *Wireless Sensor Networks: First European Workshop, EWSN 2004, Berlin, Germany, January 19–21, 2004, proceedings*. New York: Springer-Verlag, 2004.
5. T. Yan et al., "Design and optimization of distributed sensing coverage in wireless sensor networks," *ACM Transactions on Embedded Computing Systems* 7, 2008.
6. I.F. Akyildiz et al., "Wireless multimedia sensor networks: A survey," *IEEE Wireless Communications* 14, December 2007, pp. 32–39.
7. V. Rajaravivarma et al., "An overview of wireless sensor network and applications," *Proceedings of the 35th Southeastern Symposium on System Theory*, 2003, pp. 432–436.
8. A. Goldsmith, *Wireless Communications*. Cambridge: Cambridge University Press, 2005.
9. T.S. Rappaport, *Wireless Communications: Principles and Practice*, 2nd ed. Upper Saddle River, Prentice Hall PTR, 2002.
10. J.G. Proakis and M. Salehi, *Digital Communications*, 5th ed. Boston: McGraw-Hill, 2008.
11. M.K. Simon and M.-S. Alouini, *Digital Communication over Fading Channels*, 2nd ed. Hoboken, Wiley-Interscience, 2005.
12. M.S. Alouini and A.J. Goldsmith, "Adaptive modulation over Nakagami fading channels," *Wireless Personal Communications* 13, May 2000, pp. 119–143.

13. H. Matsuoka et al., "Adaptive modulation system with variable coding rate concatenated code for high quality multi-media communication systems," *1996 IEEE 46th Vehicular Technology Conference, Proceedings, Vols. 1–3,* 1996, pp. 487–491.

14. H. Alasady and M. Ibnkahla, "Adaptive modulation over nonlinear time-varying channels," *European Transactions on Telecommunications* 18, November 2007, pp. 685–692.

15. A.J. Goldsmith and P.P. Varaiya, "Capacity of fading channels with channel side information," *IEEE Transactions on Information Theory* 43, November 1997, pp. 1986–1992.

16. J.K. Cavers, "Variable-rate transmission for Rayleigh fading channels," *IEEE Transactions on Communications* Co20, 1972, p. 15.

17. J.F. Hayes, "Adaptive feedback communications," *IEEE Transactions on Communication Technology* Co16, 1968, p. 29.

18. S.T. Chung and A.J. Goldsmith, "Degrees of freedom in adaptive modulation: A unified view," *IEEE Transactions on Communications* 49, September 2001, pp. 1561–1571.

19. W.T. Webb and R. Steele, "Variable-rate QAM for mobile radio," *IEEE Transactions on Communications* 43, July 1995, pp. 2223–2230.

20. A.J. Goldsmith and S.G. Chua, "Variable-rate variable-power MQAM for fading channels," *IEEE Transactions on Communications* 45, October 1997, pp. 1218–1230.

21. D.L. Goeckel, "Adaptive coding for time-varying channels using outdated fading estimates," *IEEE Transactions on Communications* 47, June 1999, pp. 844–-855.

22. J. Hagenauer, "Rate-compatible punctured convolutional-codes (RCPC codes) and their applications," *IEEE Transactions on Communications* 36, April 1988, pp. 389–400.

23. M. Rice and S.B. Wicker, "Adaptive error control for slowly varying channels," *IEEE Transactions on Communications* 42, February–April 1994, pp. 917–926.

24. B. Vucetic, "An adaptive coding scheme for time-varying channels," *IEEE Transactions on Communications* 39, May 1991, pp. 653–663.

25. T. Ue et al., "Symbol rate and modulation level-controlled adaptive modulation TDMA TDD system for high-bit-rate wireless data transmission," *IEEE Transactions on Vehicular Technology* 47, November 1998, pp. 1134–1147.

26. A. Ghosh et al., "Broadband wireless access with WiMax/802.16: Current performance benchmarks and future potential," *IEEE Communications Magazine* 43, February 2005, pp. 129–136.

27. Q.W. Liu et al., "Cross-layer combining of adaptive modulation and coding with truncated ARQ over wireless links," *IEEE Transactions on Wireless Communications* 3, September 2004, pp. 1746–1755.

28. F. Peng et al., "Adaptive modulation and coding for IEEE 802.11n," *2007 IEEE Wireless Communications & Networking Conference, Vols. 1–9,* 2007, pp. 657–662.

29. J.L.C. Wu et al., "An adaptive multirate IEEE 802.11 wireless LAN," *15th International Conference on Information Networking, Proceedings,* 2001, pp. 411–418.

30. T.H. Chan, M. Hamdi, C.Y. Cheung, and M. Ma, "A link adaptation algorithm in MIMO-based WiMAX systems," *Journal of Communications* 2, August 2007, pp. 16–24.

31. B. Classon et al., "Channel coding for 4G systems with adaptive modulation and coding," *IEEE Wireless Communications* 9, April 2002, pp. 8–13.

32. A. Furuskar et al., "EDGE: Enhanced data rates for GSM and TDMA/136 evolution," *IEEE Personal Communications* 6, June 1999, pp. 56–66.
33. S. Nanda et al., "Adaptation techniques in wireless packet data services," *IEEE Communications Magazine* 38, January 2000, pp. 54–64.
34. S. Otsuki et al., "Square-QAM adaptive modulation/TDMA/TDD systems using modulation level estimation with Walsh-function," *Electronics Letters* 31, February 2, 1995, pp. 169–171.
35. A.Y. Alemdar, "Link adaptation for energy constrained networks," MSc thesis, Department of Electrical and Computer Engineering, Queen's University, Kingston, 2008.
36. A. Chandrakasan et al., "Design considerations for distributed microsensor systems," *Proceedings of the IEEE 1999 Custom Integrated Circuits Conference*, 1999, pp. 279–286.
37. J.M. Rabaey et al., "PicoRadio supports ad hoc ultra-low power wireless networking," *Computer* 33, July 2000, p. 42.
38. B.M. Sadler, "Fundamentals of energy-constrained sensor network systems," *IEEE Aerospace and Electronic Systems Magazine* 20, August 2005, pp. 17–35.
39. Y.W. Hong et al., "Cooperative communications in resource-constrained wireless networks," *IEEE Signal Processing Magazine* 24, May 2007, pp. 47–57.
40. Y.T. Hou et al., "On energy provisioning and relay node placement for wireless sensor networks," *IEEE Transactions on Wireless Communications* 4, September 2005, pp. 2579–2590.
41. R. Min et al., "Energy-centric enabling technologies for wireless sensor networks," *IEEE Wireless Communications* 9, August 2002, pp. 28–39.
42. S.C. Cripps, *Advanced Techniques in RF Power Amplifier Design*. Boston: Artech House, 2002.
43. K. Akkaya and M.F. Younis, "Survey on routing protocols for wireless sensor networks," *Ad Hoc Networks* 3, 2005, pp. 325–349.
44. K. Sohrabi et al., "Protocols for self-organization of a wireless sensor network," *IEEE Personal Communications* 7, October 2000, pp. 16–27.
45. C. Intanagonwiwat et al., "Directed diffusion for wireless sensor networking," *IEEE Transactions on Networking* 11, February 2003, pp. 2–16.
46. J. Van Greunen et al., "Adaptive sleep discipline for energy conservation and robustness in dense sensor networks," presented at the 2004 IEEE International Conference on Communications, 2004.
47. D.P. Van Greunen J., A. Bonivento, J. Rabaey, K. Ramchandran, A.S. Vincentelli, "Adaptive sleep discipline for energy conservation and robustness in dense sensor networks," presented at the IEEE International Conference on Communications, 2004.
48. J.M. Rabaey et al., "PicoRadio supports ad hoc ultra-low power wireless networking," *Computer* 33, July 2000, p. 42.
49. J.M. Kahn et al., "Emerging challenges: Mobile networking for 'Smart Dust,'" *Journal of Communications and Networks* 2, September 2000, pp. 188–196.
50. P. Agrawal and N. Patwari, "Correlated link shadow fading in multi-hop wireless networks," *IEEE Transactions on Wireless Communications* 8, August 2009 pp. 4024–4036.
51. N. Patwari and P. Agrawal, "Effects of correlated shadowing: Connectivity, localization, and RF tomography," *2008 International Conference on Information Processing in Sensor Networks, Proceedings*, 2008, pp. 82–93.

52. M. Ilyas and I. Mahgoub, *Handbook of Sensor Networks: Compact Wireless and Wired Sensing Systems*. Boca Raton, Fla.: CRC Press, 2005.
53. J.N. Al-Karaki and A.E. Kamal, "Routing techniques in wireless sensor networks: A survey," *IEEE Wireless Communications* 11, December 2004, pp. 6–28.
54. A.S. Tanenbaum, *Computer Networks*, 4th ed. Upper Saddle River, NJ: Prentice Hall PTR, 2003.
55. V.K.N. Lau and Y.-K.R. Kwok, *Channel-Adaptive Technologies and Cross-Layer Designs for Wireless Systems with Multiple Antennas: Theory and Applications*. Hoboken, John Wiley, 2006.
56. D. Wagner and R. Wattenhofer, *Algorithms for Sensor and Ad Hoc Networks: Advanced Lectures*. Berlin: Springer, 2007.
57. W. Ye et al., "An energy-efficient MAC protocol for wireless sensor networks," *IEEE Infocom 2002: The Conference on Computer Communications, Vols. 1–3, Proceedings*, 2002, pp. 1567–1576.
58. D.P.J. Van Greunen, A. Bonivento, J. Rabaey, K. Ramchandran, A.S. Vincentelli, "Adaptive sleep discipline for energy conservation and robustness in dense sensor networks," presented at the IEEE International Conference on Communications, 2004.
59. R. Jurdak et al., "Radio sleep mode optimization in wireless sensor networks," *IEEE Transactions on Mobile Computing* 9, July 2010, pp. 955–968.
60. C. Park et al., "Battery discharge characteristics of wireless sensor nodes: An experimental analysis," *2005 Second Annual IEEE Communications Society Conference on Sensor and Ad Hoc Communications and Networks*, 2005, pp. 430–440.
61. E. Casilari et al., "Modeling of current consumption in 802.15.4/ZigBee sensor motes," *Sensors* 10, June 2010, pp. 5443–5468.
62. M. Alnuaimi et al., "Performance evaluation of IEEE 802.15A physical layer using MATLAB/Simulink," *2006 Innovations in Information Technology*, 2006, pp. 156–160.
63. W.T.H. Woon and T.C. Wan, "Performance evaluation of IEEE 802.15.4 wireless multi-hop networks: simulation and testbed approach," *International Journal of Ad Hoc and Ubiquitous Computing* 3, 2008, pp. 57–66.
64. F.M. Al-Turjman et al., "Connectivity optimization for wireless sensor networks applied to forest monitoring," *2009 IEEE International Conference on Communications, Vols. 1–8*, 2009, pp. 285–290.
65. X. Zhao, *Energy Constrained Link Adaptation for Multi-hop Relay Networks*, MSc. diss., Queen's University, Canada, 2010.
66. X. Zhao, E. Bdira, and M. Ibnkahla, "Adaptive modulation and MAC protocols for wireless sensor networks," submitted for publication, *International Journal of Distributed Sensor Networks*, submitted for publication, December 2011.

4

Cross-Layer Approaches to QoS Routing in Wireless Multihop Networks

4.1 Introduction

Popular categories of multihop networks include wireless sensor network (WSN), mobile ad hoc network (MANET), wireless mesh network (WMN), and vehicular ad hoc network (VANET). In WSNs, sensor nodes are deployed in the target area to measure specific attributes, such as temperature or pressure, and relay the measured information to a base station for processing. Sensor nodes can be deployed in large numbers and can operate autonomously without human intervention. Nodes in MANETs typically have sufficient processing and networking capabilities and can connect together autonomously and run a variety of applications. In WMNs, nodes have similar capabilities as MANETs, but some infrastructure is typically utilized, mainly to deliver Internet services to a large number of wireless devices. On the other hand, in VANETs vehicles are equipped with transceivers that may be used to exchange information such as traffic intensity and collision warnings, or for regular data communication.

In addition to the immense popularity of multihop networks, wireless users are increasingly demanding the support of more challenging applications, such as video streaming and voice-over Internet protocol (VoIP). Such applications require the support of certain quality of service (QoS) parameters, such as bandwidth, end-to-end delay, packet loss ratio (PLR), and jitter, for efficient operation. Yet despite extensive research efforts, the goal of guaranteeing a satisfactory QoS level over multihop networks remains elusive. Generally speaking, hard QoS, where parameters are strictly guaranteed for the entire duration of the communication session, is very difficult to achieve in multihop networks due to their highly dynamic nature. Instead, researchers strive to improve the support of soft QoS. This means that, although parameters may be ensured when the communication session starts, there might be periods when the required parameters violate their thresholds.

There is no general framework for QoS support over all types of multihop networks. Different types of multihop networks impose different challenges and constraints that have to be considered during protocol design [1] [2] [3]. In order to accurately study the challenges behind QoS support, and due to the pressing demand for QoS applications, we present a survey of the most up-to-date efforts on routing design that target the support of QoS applications over different types of multihop networks. We have chosen to address four types of multihop networks: MANET, WMN, WSN, and VANET. Figure 4.1 shows the multihop networks that are considered in this chapter. These are the ones with the most potential and the target of most researchers. In addition, they provide insights into the support of QoS over other types of multihop networks, such as cellular multihop networks.

One of the key aspects of this chapter is that we mainly consider routing protocols that employ cross-layer design, which has now become a fundamental design concept among wireless researchers. Cross-layer design is the process by which parameters are exchanged between layers for the purpose of enhancing the overall performance of the network.

The reason cross-layer design has been gaining increasing attention from the research community is that, in reality, changes made in one layer can have a profound effect on other layers in the network. For example, power adaptation can lead to the creation or deletion of links, thus changing the entire network topology, which can have a significant effect on routing decisions at the network layer. Scheduling transmission is done at the medium access control (MAC) layer and will affect the level of interference at the

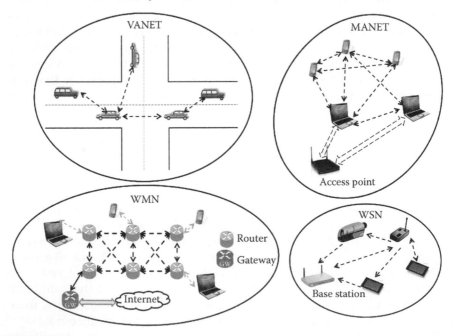

FIGURE 4.1
Examples of wireless multihop networks.

physical (PHY) layer. Even flow control at the transport layer will affect the overall level of congestion in the network [4]. Cross-layer design can thus have a profound impact on QoS support over multihop networks. Examples of cross-layer design include combining routing at the network layer with scheduling and call admission at the MAC layer and power adaptation at the PHY layer in order to control interference, or combining channel assignment at the MAC layer with adaptive modulation at the PHY layer to optimize wireless resources over each link.

This chapter considers QoS support from such a broad perspective by reviewing several network types and the inherent challenges within each. Such a survey is important in illustrating why there is no general framework yet for QoS in all multihop networks and shows the conditions where different routing protocols may be suitable. We also discuss the challenges that need to be considered if such a generic framework is to be designed or if interoperability between different networks is to be realized.

The remaining sections of this chapter are organized as follows: Section 4.2 highlights the main design considerations that should be included when designing a QoS routing protocol. In Section 4.3 some cross-layered approaches that represent the state of the art in QoS routing design are surveyed for different types of multihop networks. In Section 4.4, the surveyed protocols are compared and their advantages and disadvantages highlighted, while Section 4.5 presents open challenges and future research directions that promise to improve network performance. Finally, Section 4.6 presents some concluding remarks to the chapter.

4.2 Design Challenges and Considerations

There are multiple factors that have to be considered during the design of a QoS routing protocol. These factors largely depend on the applications to be supported and the nature of the network itself. This section is devoted to highlighting some of the important challenges and considerations of QoS routing protocol design.

4.2.1 QoS Metrics

Different applications require the support of different resources over the wireless channel. Generally speaking, in multihop networks, these resources have to be supported end to end—that is, from source to destination—possibly across multiple intermediate nodes. The main metrics that most applications require for satisfactory performance are [5]:

- *Minimum Throughput:* This might not necessarily refer only to the PHY layer resource of bandwidth or raw data rate but may also refer to an average rate of successful packet delivery that needs to be guaranteed at the MAC layer. Throughput is primarily important in video and voice applications or any application where a large amount of information needs to be transported.

- *Maximum Delay:* This is mainly composed of propagation delays, plus queuing and processing delays. This metric is particularly important in VANETs, due to the critical nature of relaying information about safety.

- *Maximum Delay Jitter:* This is usually defined as the difference between the delay upper bound and the minimum possible delay (which is usually defined by the propagation time and packet length). Jitter is important in applications where a constant flow of information is required, such as in video streaming.

- *Maximum Packet Loss Ratio (PLR):* This is defined as the maximum ratio of packets that can be lost before significantly degrading the integrity of the transmitted data. High PLR can be caused by congestion or bad channel conditions and is of importance to most applications.

4.2.2 Design Challenges

The protocol design challenges depend mainly on the nature of the network. Therefore, we classify the challenges according to the network type [6].

- *MANET Challenges:* Nodes in MANETs typically have sufficient energy and processing capabilities. However, they suffer from the lack of centralized control. Tasks such as resource management, admission control, and scheduling have to be performed in a distributed way, which may result in suboptimal performance. Interference between nodes is also a major concern and is caused by contention for the wireless medium. It can be classified into interflow and intraflow interference. Interflow interference is the interference between different flows, while intraflow interference is the interference between multiple nodes on the same flow. Interference also has to be managed in a distributed way. Cross-layered approaches for MANETs include joint optimization of power, data rate, scheduling, and admission control tasks.

- *WMN Challenges:* Nodes in WMNs also have sufficient energy and processing capabilities and share some of the challenges of MANETs, particularly interference. However, the presence of some infrastructure means that some tasks can be managed in a centralized way. Scalability is a major challenge in WMNs due to their typical large size. Protocols have to be able to consider a large number of nodes with minimum possible complexity. Topology control is also important in order to avoid congested areas, resulting from the heavy traffic load typically present in WMNs, which is why joint design of routing and link scheduling is a popular cross-layer research direction. Network planning and deployment can also be critical to the overall performance, since they affect load balancing and future network expansion.

- *VANET Challenges:* High mobility is the main cause for most challenges in VANETs [7]. This causes the topology to be highly dynamic, where links may not be valid for more than a few seconds. It also means that the network may suffer from frequent partitions, where subsets of nodes may become isolated. Different routing considerations also have to be included in the routing protocol. For example, the mobility pattern of vehicles is deterministic because vehicles are bound by roads. Also, the protocol has to consider different communication environments that may be present in different roads (for example, communication on a highway has different requirements than communication in downtown). Delay is usually the main QoS metric requested by VANET applications and has to be the main concern of the routing protocol. Most cross-layer design techniques for VANETs utilize GPS devices to aid in routing. Nodes in VANETs also have sufficient energy, and storage capabilities and restrictions on the size of the devices are usually relaxed. This means that larger antennas can be employed in VANETs to improve performance.

- *WSN Challenges:* Due to the strict requirements on the size of sensor nodes, they usually have limited energy, storage, and processing capabilities. The main challenge in WSN protocol design is energy efficiency. WSN networks are usually required to remain operational for several months without human intervention, using the small batteries on board the sensor devices. Sensor networks also operate at relatively low data rates. These characteristics make QoS for WSNs a highly challenging task. QoS routing protocols have to be simple (avoiding operations with high complexity) and should avoid wasted transmissions and unnecessary overhead, since transmissions are considered the largest consumer of energy in the network. Load balancing also needs to be considered in order to avoid the overuse of any subset of nodes. Cross-layer protocols for WSNs often consider remaining battery levels and may consider parameters such as contention from the MAC layer or geographical information from the PHY layer during routing, in order to ensure energy efficiency.

Table 4.1 summarizes these challenges and highlights the main cross-layer directions that address those challenges.

TABLE 4.1

Multihop Network Challenges and Cross-Layer Solutions

	MANET	WMN	VANET	WSN
Considerations and Challenges	• Lack of centralized control • Interflow and intraflow interference • Distributed resource management	• Interference management • Scalability • Link scheduling • Topology control • High traffic load	• Dynamic topology • Network partitions • Unknown network size • Deterministic mobility pattern • Different routing requirements	• Limited energy • Limited storage and processing • Limited bandwidth
Popular Cross-Layer Approaches	• Joint routing and admission control • Joint routing and power control • Joint routing and scheduling	• Joint routing and scheduling • Joint routing and admission control • Joint routing and topology control	• Geographical routing • Joint routing and power control	• Joint routing and topology control • Joint routing and scheduling • Geographical routing

4.2.3 Network Resources and Performance Metrics

There are many ways to evaluate the performance of a routing protocol. This can be done by directly measuring the QoS metrics required by the application, or through other means that can reflect how the protocol operates. In this section the main network resources available at each layer are presented, and the performance metrics typically employed at each layer are discussed [5] [6].

- *PHY Layer Resources and Performance Metrics:* Resources at this layer include available bandwidth or channel capacity (which largely depend on channel conditions and interference) and remaining battery power. Bandwidth is often measured using the average rate of channel utilization (how long the channel remains busy) or using interference models. These models can be based on signal-to-interference plus noise ratio (SINR) measurements or the link conflict graph, which identifies the links that can operate simultaneously. In addition, node locations and mobility levels are measured at the PHY layer and can be important metrics for routing decisions. End-to-end delay, bit error rate (BER), SINR, and rate of battery utilization are often used to evaluate the performance of the PHY layer.

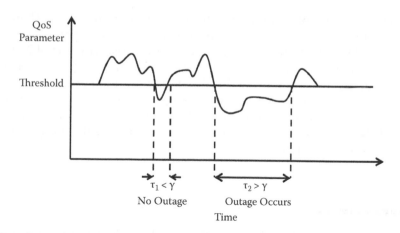

FIGURE 4.2
Outage of a QoS parameter.

- *MAC Layer Resources and Performance Metrics:* Buffer space is one of the most important MAC layer resources since it has a profound effect on delay and congestion levels. Link throughput (per-node throughput) is also measured at the MAC layer, since this layer is responsible for determining the success or failure of packet reception. Probability of contention is often considered by some routing protocols, and it is typically measured by counting the number of packets lost over a period of time. Link reliability, stability, and lifetime are also measured at the MAC layer and have a direct effect on PLR.

- *Network Layer Resources and Performance Metrics:* Congestion is typically measured at the network layer, since it requires a view of the overall network resources. Similarly, aggregate throughput (total network throughput) and network lifetime also require a network view and should be measured at the network layer.

- *Upper-Layer Performance Metrics:* QoS routing protocols are often evaluated from the perspective of the application itself. For example, session blocking/dropping ratios and false rejection ratios are some of the important methods of evaluating admission control protocols. A large session-blocking ratio implies that the protocol may be too conservative, while a large session-dropping ratio implies that the protocol may be admitting too many sessions. Measuring the performance of routing protocols from the user's or the application's perspective can be of great importance, especially for marketing or business purposes.

Finally, it is worth noting that the dynamic nature of all multihop networks and the large number of considerations and interactions that may affect network performance usually make it difficult to guarantee QoS parameters for the entire duration of the communication session. Therefore, in evaluating the performance of any protocol, it is more practical to consider the outage of a certain metric. To illustrate, consider Figure 4.2. As the figure illustrates, the application defines a certain duration, γ, such that if the QoS parameter drops below the threshold for a duration greater than γ, the integrity of the application is lost and outage occurs. However, most applications may tolerate a period during which the QoS parameter is allowed to violate the threshold while still being able to preserve the integrity of the transmitted information.

4.3 Taxonomy of QoS Routing Protocols in Multihop Networks

In this section some recent protocols that target QoS support in different types of multihop networks are presented. The protocols presented in this section are classified according to the type of network they address.

4.3.1 QoS Routing in MANETs

MANETs are among the most popular types of multihop networks, owed mainly to the widespread implementation of the IEEE standard 802.11 (Wi-Fi) on wireless devices. Early work on routing protocols for MANETs focused on the task of route discovery, and paths were chosen by utilizing routing metrics. In cross-layer design, metrics can be combined between layers in order to improve performance. Protocols such as ad hoc on-demand distance vector (AODV) routing [8], dynamic source routing (DSR), and destination-sequence distance vector (DSDV) routing [9] simply choose the shortest path from a source node to a destination node. The expected transmission count (ETX) metric [10] chooses paths that maximize packet delivery ratio by utilizing information from the MAC layer about the number of successfully transmitted and received packets. On the other hand, the weighted cumulative expected transmission time (WCETT) metric [11] chooses paths with the highest bandwidth by utilizing information from the PHY layer about channel status.

However, recent research [12] shows that combining metrics might not lead to the best results in the context of QoS support. Adding, multiplying, or dividing several metrics produces a cost that only reflects the combination of these metrics but does not reflect each individual metric by itself. This can be illustrated using Figure 4.3, which shows three possible paths between a source node S and destination D. In Figure 4.3(a), metrics of delay and bandwidth are aggregated into a single cost metric. Thus, the path with the least cost will be chosen every time regardless of what the QoS application requires. In Figure 4.3(b), each QoS parameter is computed separately for each path, which provides more information for routing decisions. For

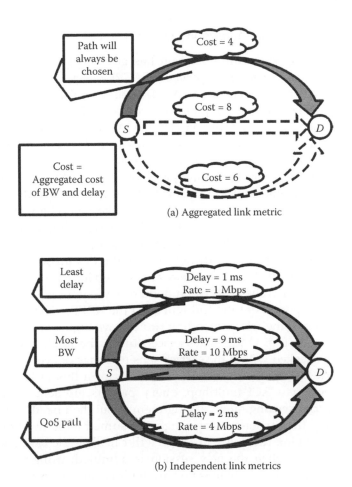

(a) Aggregated link metric

(b) Independent link metrics

FIGURE 4.3
Route selection in multihop networks.

example, if the application requires a path with minimum delay, then the first path will be chosen, while if a path is required that supports both low delay and high bandwidth, the third path will be chosen. Thus, computing routing metrics separately will lead to better QoS support.

In addition to route discovery and accurate resource estimation, call admission protocols at the MAC layer are of particular importance in MANETs in order to control transmissions according to the available wireless resources and reduce interference, which is a key challenge for QoS support. The main challenge in call admission is to consider the available resources within the carrier-sensing range of each node and not just within the transmission or local range. This is necessary to ensure that newly admitted sessions do not degrade the performance of currently active sessions.

4.3.1.1 Contention-Aware Admission Control Protocol

Combining routing with call admission has been explored in [13], where *contention-aware admission control protocol* (CACP) was proposed. CACP is a cross-layered scheme between network, MAC, and PHY layers, where interference within the carrier-sensing range is considered during resource estimation and used for call admission at the MAC layer. Admission of new flows is divided into two phases. The first phase includes route discovery and estimation of local resources (within the transmission range) at every node, which is done using a source routing protocol such as DSR. The network is flooded with route request (RREQ) packets. Upon receiving a RREQ packet, every intermediate node measures its local available bandwidth. This is done by monitoring the busy time of the channel. A node will only forward the RREQ packet if its local bandwidth is greater than the requirements of the session to be admitted. As the RREQ reaches the destination, it will only send a route reply (RREP) packet if there are enough local resources at every node along the path. If the destination node receives multiple RREQ, it will cache them for a period of time and will send the RREP on one route (more RREP can be sent if their corresponding routes have higher bandwidth).

The second phase of CACP takes place during the RREP phase, where resources are checked within the carrier-sensing range of every node along the route. CACP proposes three methods to perform this task. In the first method, known as CACP-multihop, query packets are sent to all nodes within the carrier-sensing range, asking them about their bandwidth usage. In this method, the carrier-sensing range is assumed to be two hops wide. Every node receiving the query message will calculate the bandwidth that will be consumed when the new session is admitted, using information within the query packet. If there is no bandwidth available to accommodate the new session, a reply will be sent and the route will be rejected. If the RREP reaches the source node, it will know that all the intermediate nodes have sufficient resources and that the new admitted route will not violate the requirements of other active sessions.

The second call admission method proposed by CACP is known as CACP-power. In this method the query packet is sent at a higher transmit power, capable of reaching all nodes within the carrier-sensing range. Thus the packet only needs one hop to reach all nodes to be queried. As in CACP-multihop, admission will only take place if the RREP propagates back to the source node. CACP-power requires less overhead than CACP-multihop but needs some hardware modifications to support the increased transmit power. In the third algorithm proposed by CACP, known as CACP-CS, admission is performed at node *A* by monitoring the available bandwidth using a sensing threshold that covers the carrier-sensing ranges of all nodes within the carrier-sensing range of *A* (this threshold is known as neighbor carrier-sensing range). This method requires no extra overhead or hardware

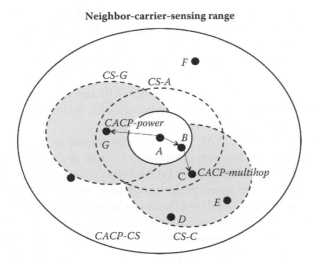

FIGURE 4.4
Admission algorithms in CACP.

modifications but is less accurate than CACP-multihop and CACP-power. The three admission algorithms are illustrated in Figure 4.4.

Although a network would only use one of the three proposed methods, they are all illustrated in Figure 4.4 for comparison. CS-x denotes the carrier-sensing range of node x. Thus, the neighbor-carrier-sensing range of A covers CS-$\{A, B, C, G\}$. The main disadvantage of CACP is that it is overly conservative. The algorithms may include nodes outside the carrier-sensing range of the nodes to be admitted. However, to the best our knowledge, CACP was the first protocol to consider resources within the carrier-sensing range during admission, which is why it was explained in detail in this section.

A modification to CACP-CS was proposed in [14], which we label routing and admission control (RAC), by assuming that transmissions within the neighbor-carrier-sensing range are independent from transmissions within the carrier-sensing range of the node to be admitted. Thus the estimated available bandwidth is not as conservative as in CACP-CS. RAC also assumes that different links can support different data rates, and uses a route discovery method that allows for quick route recovery, which is not available in CACP.

4.3.1.2 Adaptive Admission Control

In order to address the conservative nature of CACP and provide more accurate call admission, a cross-layered protocol named *adaptive admission control* (AAC) was proposed [15]. This protocol utilizes information from the MAC and PHY layers to provide proactive resource discovery and perform admission control at the MAC layer. Periodic HELLO packets are used to broadcast the available bandwidth measured at every node (by monitoring the busy time of the channel as in CACP). Although HELLO packets are transmitted only to the one-hop range of every node, bandwidth information is aggregated in every HELLO message so that each node will learn the bandwidth information of other nodes within the carrier-sensing range. Therefore, when a session needs to be admitted, nodes will already know the available bandwidth at every hop. The available bandwidth is defined in AAC as the minimum bandwidth of all interfering nodes.

The routing protocol of AAC also includes means of dealing with congestion and mobility. Whenever congestion occurs at a particular node (for example, if it is being used to forward data in multiple sessions) or if mobility causes a node to enter the interference range of an ongoing session, the node where QoS thresholds are violated chooses the session consuming the most resources and informs the source node of that session to pause transmission for a while. This frees up resources for other sessions and minimizes the number of sessions that will be dropped. However, the session that is paused might not be admitted again if congestion does not decrease.

4.3.1.3 Interference-Aware QoS Routing

Another direction was pursued in [16], where *interference-aware QoS routing* (IQRouting) was proposed. Cross-layer interactions are utilized between the network and PHY layers in order to formulate the link conflict graph, which portrays the interference among links in the network, and perform admission control at the MAC layer. Nodes in a link conflict graph represent links of the network, and edges of the graph connect links that are interfering with each other (within the interference range). From graph theory, a complete subgraph is called a clique, and so all links in a clique interfere with each other. The concepts of conflict graphs and cliques can be illustrated using Figure 4.5.

From Figure 4.5, links in {A, B, F}, {A, B, E}, {B, E, F}, and {A, E, B, F} are all cliques, since they all interfere with each other. IQRouting broadcasts periodic control messages throughout the network to discover the network

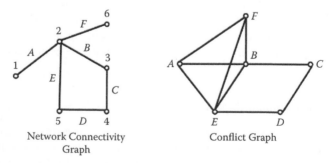

Network Connectivity
Graph

Conflict Graph

FIGURE 4.5
Interference among links modeled using a conflict graph.

topology and cliques. Information about bandwidth consumption is periodically broadcasted as well. A source routing protocol is used to discover several candidate paths, and the resources of each path are tested to find the best one. Probe packets are sent over each candidate path to check for clique constraints. The accumulated bandwidth of the path is defined as the minimum bandwidth of all its links. When the probe packets reach the destination they are compared, and the one with the largest bandwidth is chosen. IQRouting requires significant overhead to maintain control information, establish the link conflict graph, and perform admission, and therefore will not perform well under high-mobility scenarios.

4.3.1.4 Interference-Based Fair Call Admission Control

The cross-layer routing protocol proposed in [17], known as *interference-based fair call admission control* (iCAC), is one of the few protocols that consider fairness while performing call admission at the MAC layer. The channel busy time is monitored at the PHY layer to determine the available resources at every node. The main advantage of iCAC is that nodes divide bandwidth evenly among all nodes within their carrier-sensing range. Periodic HELLO packets are used by nodes in order to maintain neighbor tables and announce the bandwidth usage. However, the HELLO packets are only broadcasted every five seconds to reduce overhead. When a node needs to admit a new flow, it uses information from the HELLO packets to determine if there is enough bandwidth available and informs all nodes within its carrier-sensing range that are currently engaged in communication sessions of their bandwidth assignments. iCAC assumes that all nodes have a reserved transmit power capable of reaching all nodes within the carrier-sensing range.

Another important contribution made in [17] is how iCAC considers the channel to be busy. Rather than considering the channel to be busy only when the received power is greater than the carrier-sensing threshold, the work in [17] identifies that this is in fact the upper bound on channel busy time. The lower bound on a node's busy time is calculated by considering the channel busy only when the node is transmitting or receiving. iCAC uses those bounds to determine if the interferers are located inside a node's transmitting range or within its carrier-sensing range. Every node senses the medium using the receiving threshold (the power level below which a packet cannot be received) and the carrier-sensing threshold. If the channel is considered idler using the receiving threshold than the carrier-sensing threshold, then more interferers are located within the carrier-sensing range than within the transmitting range.

It is interesting to point out that ideas for QoS support proposed for MANETs can be adopted for WMNs, and vice versa, due to the similarities between the two networks. Nodes in both networks have sufficient energy and networking capabilities, and thus can support challenging applications. Ideas for QoS support for WMNs are presented in the next section.

4.3.2 QoS Routing in WMNs

Traffic demand in WMNs can be high, similar to the case of MANETs. In fact, early protocols proposed for MANETs, such as AODV and WCETT, have also been used in WMNs. However, the key difference between WMNs and MANETs is that WMNs rely on some infrastructure that can perform some centralized management operations, which can be highly valuable in controlling interference. WMNs are also typically larger in size than MANETs and target the support of heavier traffic demands. Mobility in WMNs is also typically lower than MANETs. In this section we focus on those protocols that were designed specifically for WMNs.

4.3.2.1 QoS Routing and Distributed Scheduling

Due to the high traffic demands, combining routing at the network layer with scheduling at the MAC layer is a highly popular cross-layer approach in WMNs in order to avoid bottlenecks and ensure the desired QoS levels. In [18] a framework for *QoS routing and distributed scheduling* (QRDS) was proposed that utilizes extensive cross-layer interactions between network, MAC and PHY layers to support multiple QoS metrics. Since WMNs target the support of a wide variety of applications, QRDS includes specifications for supporting the QoS metrics of throughput, delay, and PLR. A utility function that avoids bottlenecks is defined based on a dissatisfaction ratio for each QoS metric. These utility functions are defined as:

- *Delay dissatisfaction ratio* is the ratio between the actual delay measured over a path and the required QoS delay.
- *Throughput dissatisfaction ratio* is the ratio between the required QoS throughput and the measured bottleneck throughput (minimum throughput of all links) of the path.
- *PLR dissatisfaction ratio* is defined as the ratio between the product of the one-hop PLRs of all links of the path and the required QoS PLR.

Resources are estimated at the PHY layer during route discovery, and the route that minimizes the utility function is chosen by the network layer. In order to guarantee QoS parameters, resource reservation is utilized at every hop during route discovery, and a reservation margin is used to account for estimation errors. The proposed scheduler at the MAC layer also prioritizes flows whose parameters are close to being violated. A utility metric, similar to the dissatisfaction ratio, is also defined at the MAC layer in case the node is responsible for forwarding packets for several active sessions. Every time a packet needs to be forwarded, the scheduler will activate the session whose QoS metric is closest to its threshold. This ensures that the scheduler prevents the violation of QoS metrics of all sessions. Simulation results show that this algorithm outperforms the classic AODV algorithm with round-robin scheduling.

4.3.2.2 Robust Routing and Scheduling

The problem of inaccuracy of traffic information was addressed in [19], where a *robust routing and scheduling* (RRS) algorithm was proposed. The algorithm mainly targets networks where a centralized controller is available and represents a framework for cross-layer design between network, MAC, and PHY layers. The use of a centralized controller is justified by the fact that the protocol achieves worst-case optimal performance even under varying traffic conditions. An interference model is utilized at the PHY layer that identifies links that can operate simultaneously, based on either the link conflict graph or SINR measurements.

The centralized controller receives flow requests and passes a schedule for transmissions to the MAC layer at all nodes, after solving a linear programming problem (LPP) that minimizes the maximum congestion level for all flows. A traffic matrix (TM) is formulated that contains all the traffic requests, and two variations of RRS are proposed. In the first variant, no traffic knowledge is provided, and all entries in the TM can take values between 0 and ∞. In this case the algorithm considers all possible TMs and finds a routing-scheduling pair that achieves worst-case minimum congestion. In the second variant of RRS, where some traffic knowledge is available, some entries in the TM may be known. This introduces more constraints to the LPP and restricts the search space. The formulated LPP is then used to find the optimal routing-scheduling pair. The algorithm also minimizes overhead since the schedule does not have to be recomputed and redistributed as long the actual traffic is within the estimated range. In case delay is also a required metric, a constraint is added to the LPP in order to limit the number of hops taken from source to destination. By limiting the hop-count, RRS guarantees an upper bound on the delay that may be incurred.

4.3.2.3 Admission Control Algorithm Using Accurate Bandwidth Estimation

A different direction was proposed in [20], where a comprehensive *admission control algorithm* (ACA) was proposed, based on cross-layer design between network and MAC layers. Admission control is also of particular importance in WMNs due to the expected high traffic load. The advantage of ACA is that it can be implemented in a distributed way and does not require significant overhead. Every node in the network maintains an estimate of the available channel bandwidth. This bandwidth is estimated by monitoring the channel's busyness ratio at the MAC layer, which reflects the rate of channel utilization. However, ACA considers the hidden node problem while measuring the available bandwidth. Traditional bandwidth estimation mechanisms assume that the probability of successful reception of request to send (RTS) packets is the same as the probability of successful reception of data packets. In case the hidden terminal problem exists, a successful RTS packet might not guarantee a successful data packet. This problem is considered in [20] by separately considering the probabilities of reception of RTS, clear to send (CTS), data, and acknowledgment (ACK) packets.

The total estimated channel bandwidth is divided between real-time (or QoS) traffic and non-real-time traffic, where the larger portion of the channel is reserved for real-time traffic. Routes are discovered using a reactive routing protocol, such as AODV. After routes have been discovered, a node wishing to initialize a real-time connection will send an admission flow request along the route to the destination, where each intermediate node will check if the requested bandwidth is greater than the estimated available bandwidth. If bandwidth is available, each intermediate node will forward the request along the path until it reaches the destination. Otherwise, a denial of service will be sent to the source node. When the destination receives the admission request, it will send an admission reply to the source node after checking its available resources.

In case a non-real-time connection is requested, a rate adjustment algorithm is used in order to avoid network congestion. In case the bandwidth available for non-real-time traffic is small, nodes are instructed to send packets at larger intervals, and the intervals are reduced as more bandwidth becomes available. This algorithm ensures fairness among nodes and ensures that real-time traffic is always prioritized over non-real-time traffic.

4.3.2.4 On Using Contention-Based vs. Contention-Free Scheduling Algorithms

Although scheduling is typically a MAC layer issue, it is frequently coupled with routing protocols as the two problems are closely related. There are mainly two classifications for scheduling algorithms: contention-based scheduling such as carrier sense multiple access with collision avoidance (CSMA-CA), and contention-free scheduling such as time division multiple access (TDMA). Each of those techniques has its own advantages and disadvantages.

CSMA-CA has the advantage of being fully distributed. Every node contends for the medium and performs scheduling on its own. There is no synchronization required or node cooperation of any kind [6]. However, the main problem with CSMA-CA is that the limited transmission and sensing ranges of nodes cause interference and hidden node problems, which degrade throughput. Under heavy traffic loads, CSMA-CA is too conservative and does not perform very well; however, the mechanism remains simple and scalable. CSMA-CA is used by many IEEE standards such as 802.11 and 802.16.

On the other hand, TDMA has the advantage of being able to handle heavy traffic loads [21]. It is also able to achieve good spatial reuse and fairness among multiple nodes. However, TDMA is more suitable for networks with a centralized controller, since a transmission schedule needs to be calculated and distributed to a group of nodes. This is why TDMA is more suitable for WMNs. In distributed networks, TDMA may require significant overhead.

One key factor in forming a schedule based on TDMA is to identify links that can operate simultaneously using an interference model [21]. Early interference models, such as the link conflict graph, assumed that the coverage area of nodes take a circular disk shape, which can be inaccurate. More accurate interference models have been employed in [21] based on SINR measurements.

Thus far, the network types we have studied have limited or no mobility. Node movement can cause changes in topology and network resources. High node mobility, which is the case in VANETs, imposes strict challenges on QoS support and has to be carefully considered. The following section examines QoS support in VANETs.

4.3.3 QoS Routing in VANETs

Although nodes in VANETs may have sufficient energy and networking capabilities, high node mobility could mean that not all applications can be supported. Applications where links are required to remain active for long periods of time will probably not observe the desired QoS level, due to frequent link breakages. This is why routing protocols in VANETs have to consider link lifetime. Most routing protocols proposed for VANETs also utilize geographical information such as road maps and vehicle positions from the PHY layer. Shorter paths are generally preferred since they have a fewer number of links and are thus considered more reliable. Traffic information can also be utilized; for instance, if two vehicles are moving in the same direction, the link between them is expected to have a longer lifetime than links between vehicles moving in opposite directions.

There are mainly two modes of operation proposed for VANETs, the first being vehicle to vehicle (V2V) and the second being vehicle to infrastructure (V2I). In V2V, the vehicles directly communicate with each other, while in V2I the vehicles communicate with roadside units that provide Internet coverage and relay information between vehicles. Due to the relative novelty of VANETs as a research topic, it is envisioned that V2V will be deployed first before the technology gains sufficient maturity, at which point companies will be willing to invest in roadside units.

In [22] a study of the highest achievable QoS for VANET routing protocols was performed. This study considered the performance metrics of link lifetime, end-to-end delay, jitter, and PLR. It was shown that current routing protocols can satisfy most application requirements of delay and jitter. However, link lifetime is highly dependent on the speed of vehicles, while PLR depends on traffic intensity and the type of environment in which the signal propagates (highway or urban). This implies that the need for roadside units will be inevitable if routing is to be performed over large areas, while small areas can be covered using V2V communication.

4.3.3.1 Multihop Routing for Urban VANET

In [23] *multihop routing for urban VANET* (MURU) was proposed, based on cross-layer design between network and PHY layers. The primary goal of MURU is to find the path with the minimum probability of link breakage, which is considered the most robust path. MURU assumes that all vehicles are equipped with a GPS module. PHY layer information such as road geometry, movement trajectory, and vehicle speed are used in choosing paths. In order to address the challenge of having an unknown network size in VANETs, the source node first formulates a rectangular area between the source and the destination, using the location of both nodes and the shortest trajectory between them. This rectangular area is known as the broadcast area and is used to discover paths. Thus, when a RREQ packet is sent by the source node, only nodes within the broadcast area are allowed to forward the packet.

A routing metric named expected disconnection degree (EDD), is proposed in [23] and used to choose paths. EDD reflects the probability of path failure within a specified period of time. As the RREQ packet floods the broadcast area, intermediate nodes calculate and accumulate the EDD metric using PHY layer information of vehicle speed, link quality, vehicle trajectory, and road geometry. The path with the smallest EDD will be chosen by the destination for the upcoming session. Simulation results show that MURU is robust in supporting the QoS metric of delay. However, the disadvantage of MURU is that the rectangular area might lead to paths that are not optimum.

4.3.3.2 QoS Routing for VANETs

In [24] *QoS routing for VANETs* (GVGrid) was proposed and is also based on cross-layer design between network and PHY layers. All vehicles are assumed to be equipped with GPS devices. The network map is divided into small grids and position information is used to forward packets. As in MURU, a rectangular region is defined for RREQ transmissions. However, RREQ is not flooded through this area, but each node selects one candidate from each neighboring grid within the rectangular area and transmits the RREQ to them. This is illustrated in Figure 4.6.

Figure 4.6 shows the rectangular grid formed when a source node S wishes to discover a path to a destination node D. Each node sends a RREQ packet to nodes in its neighboring grids that are determined to be the best candidates. Each intermediate node then forwards the RREQ to nodes in its neighboring grids in the same way as the source node. When the RREQ reaches the destination, it chooses the path with the longest expected lifetime based on a formula presented in [24]. The destination node then sends a RREP along that route back to S, which can start utilizing the route. The goal is to find the route with the minimum number of intersections, streets, and traffic signals. Candidate nodes are thus chosen by utilizing PHY layer information such

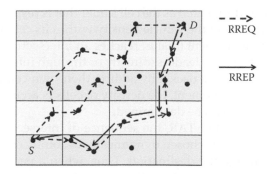

FIGURE 4.6
Route discovery process in GvGrid.

as vehicle speed and direction, as well as route characteristics (traffic lights, stop signs, and so on).

The authors of [24] also propose a route maintenance process in order to repair broken routes. The basic idea is that the initial route discovery process does not just discover a sequence of forwarding nodes from source to destination but discovers an optimum road trajectory, called driving route, on which any vehicle can provide robust and stable forwarding. Thus when a forwarding node fails, the route maintenance process attempts to find replacement nodes on the driving route, thus reducing delays in reestablishing the driving route.

4.3.3.3 Improved Greedy Traffic-Aware Routing

The cross-layered approach between network and PHY layers, known as *improved greedy traffic-aware routing* (GyTAR), was proposed in [25]. The main concept in GyTAR is to discover a sequence of intersections that are considered robust and have high vehicle densities, and use the greedy carry-and-forward mechanism to forward packets between intersections. Each vehicle is assumed to have GPS devices, and periodic HELLO packets are used to maintain neighbor tables containing position, speed, and direction of movement of neighboring vehicles. The protocol is divided into three main phases:

1. Determining the vehicle density of roads.
2. Selecting intersections that will be used to forward packets.
3. Forwarding packets between intersections using greedy forwarding.

In the first phase of GyTAR, the entire grid is divided into segments, where each segment consists of the part of the road between intersections. The goal of this phase is to discover the density of vehicles between intersections. Vehicles within each road segment are divided into location-based cells, and in each cell a node is designated as the cell leader that is responsible for maintaining information about the vehicle density within its cell. When a vehicle reaches the end of its segment (that is, reaches an intersection), it sends a cell density packet (CDP) to cell leaders (directly or through intermediate nodes), where the latter update the packet by adding their own cell density. As the CDP traverses the road segment, the density of cells is accumulated and the density of the segment can be determined.

In the second phase of GyTAR, the sequence of intersections that will be used to forward packets is chosen dynamically by jointly considering traffic density and distance to the destination. Thus, when a packet is forwarded from the source to the destination, the next intersection is chosen that is closest to the destination and has the highest traffic density. Between intersections, the third phase of GyTAR takes place, where a greedy carry-and-forward approach is used to route packets. Once a packet determines the next intersection to be reached, each node uses the neighbor table to estimate the position of its neighboring vehicles. The node that is moving toward the next intersection with the highest speed will be chosen as the next hop. In case a forwarding node cannot be found, the node simply carries the packet until the next intersection. Since forwarding is done dynamically at each intersection and on a hop-by-hop basis, route maintenance is automatically included. Simulation results in [25] show that GyTAR can achieve good performance results in terms of throughput and delay.

4.3.3.4 Stable Group-Path Routing

The routing protocol proposed in [26], which we label *stable group-path routing* (SGPR), utilizes extensive information from the PHY layer about the speed, location, and direction of movement of vehicles in order to choose stable routing paths at the network layer. The authors of [26] identify stable routing paths as those paths where links are expected to be long lived. Vehicles are divided into four mobility groups, where vehicles within the same group have the same range of velocity vectors (that is, have the same speed and direction ranges). Every group spans 90° of the Cartesian space around one of the four main directions—north, south, east, or west—as illustrated in Figure 4.7. By choosing paths where all vehicles belong to the same mobility group, stable routing can be performed and links with long lifetime can be found.

When a source node wishes to discover a path to a destination node, it broadcasts a RREQ packet. Although SGPR assumes that every vehicle is equipped with GPS systems and location maps, it does not assume that the source node knows the location of the destination. Therefore, RREQ is flooded throughout the network. The source node inserts its group ID in the RREQ packet. Upon receiving the RREQ, each intermediate node checks if it belongs to the same group, and will only rebroadcast the packet if it belongs to the same group ID as the source node. Thus, when the RREQ

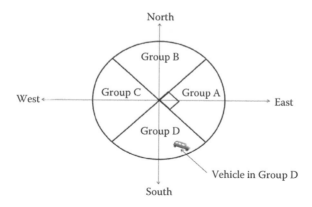

FIGURE 4.7
Mobility groups in SGPR.

reaches the destination or a node that knows a path to the destination, all the intermediate nodes along the path will belong to the same mobility group.

When the RREQ reaches the destination or a node that knows a path to the destination, a RREP packet is sent back to the source. When each intermediate node receives the RREP, it will compute a metric known as the link expiration time (LET), using information about vehicle speeds and GPS locations. Each intermediate node will check the LET field in the RREP packet and check if the newly computed metric is less than the one in the packet. If it is, it will overwrite the field with the newly calculated LET metric. This way, only the bottleneck LET is propagated back to the source, reflecting the expected lifetime of the entire path.

SGPR also employs a route recovery procedure in case a link breaks. When this happens, the node that realizes the break will check if it already knows an alternative path to the destination. If such a path is available, it will be directly utilized and the source node will be informed of the route change. If such a path cannot be found, a local recovery procedure is initiated. In case the node that realizes the break is closer to the source than the destination, then the source node will be instructed to initiate a new route discovery process. On the other hand, if the node is closer to the destination, it will send a route discovery message (RREC) in its two-hop neighborhood to find a replacement link. Nodes belonging to the same mobility group within the two-hop neighborhood will reply to the RREC, thus replacing the broken link.

4.3.3.5 Route Maintenance Mechanisms

Routing protocols for VANETs must perform path maintenance functions due to their highly dynamic nature, especially when supporting QoS applications. Traditionally, path maintenance protocols were mostly reactive, replacing nodes or links along a route after they fail. However, reactive

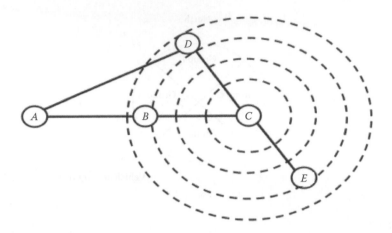

FIGURE 4.8
Route maintenance operation.

maintenance can be slow to adapt and can thus degrade the quality of QoS support. Proactive maintenance can prevent the disruption to the continuous flow of data by replacing nodes or links before QoS thresholds are violated.

In [27] a trigger-based proactive maintenance algorithm was proposed based on a reliability metric that considers node speed as well as buffer congestion. When a link is determined to be unreliable, one of the two nodes on that link will broadcast a route maintenance packet to its one-hop neighbors. To illustrate, consider Figure 4.8, which shows a multihop network where node C is transmitting packets on route ABC. When node B is determined to be unreliable and consequently link BC, node C broadcasts a special HELLO message to its one-hop neighbors, nodes D and E in the figure. Node D then determines that it is also a one-hop neighbor of A, and therefore replies to node C and replaces node B in upcoming transmissions. Simulations in [27] show that this algorithm improves throughput, delivery ratio, and end-to-end latency.

In [28], the maintenance operation is triggered when the signal strength drops below a certain threshold, mainly due to the movement of nodes outside the transmission or reception range of its neighbors. However, in [28] it is the moving node that transmits the route maintenance packet to find a replacement for itself. The authors of [28] also propose a method of wider node search if one suitable node cannot be found.

These maintenance algorithms can be utilized in any multihop network but are particularly important in VANETs. Most maintenance algorithms are based on the basic ideas illustrated in [27] and [28], where a control message is sent to perform local repair, even though maintenance metrics may vary. An alternative approach to ensure stability is to set up several routes during the discovery process and simply utilize an alternative route when one fails. This approach was followed in [29].

A completely different perspective has to be taken for QoS support in networks with strict resource constraints, as in WSNs. Limited energy and processing capabilities impose restrictions on protocol design and the nature of the applications that can be supported. Ideas for QoS support for WSNs are studied in the next section.

4.3.4 QoS Routing in WSNs

WSNs are different in nature from other types of multihop networks. They are ideal for monitoring and tracking applications, and typically operate for extended periods of time (sometimes over one year). Thus, energy is considered the scarcest resource in WSNs and routing has to be carefully designed to avoid energy waste on overhead or unnecessary transmissions.

In addition, the target of any information sent in WSNs is usually the sink node, and point-to-point communication is usually not employed. For example, data transmissions can be time driven, query driven, event driven, or a combination of these methods. Early routing protocols for WSNs include low-energy adaptive clustering hierarchy (LEACH) [30] and geographic random forwarding (GeRaF) [31]. These protocols mainly focus on choosing paths that are energy efficient. In LEACH, the network is divided into clusters, where each group of sensor nodes transmits its information to a designated node, called the cluster head. The cluster heads then communicate with each other to relay the data to the sink node. The role of the cluster head is exchanged among nodes of the same cluster in order to achieve load balancing. In GeRaF, nodes use geographical information to find the shortest path to the base station. A node with information to send simply broadcasts the data to all its active neighbors, and the neighbor that is closest to the destination will be chosen to forward the data. This process is repeated until the packet reaches the sink node.

4.3.4.1 Multipath Multispeed Routing Protocol

Although LEACH and GeRaF are considered energy-efficient protocols, they have no capabilities to accommodate QoS applications. In order to address QoS support for WSNs, *multipath multispeed protocol* (MMSPEED) was proposed in [32], and considers the MAC layer metric of reliability, as well as PHY layer metrics of delay and GPS or location information to aid in packet forwarding at the network layer. If delay is the required QoS parameter, the authors of [32] define a threshold, SetSpeed, to ensure that a maximum delay is guaranteed across the network. Every node maintains an estimate of the delay needed to transmit to each of its active neighbors (including queuing, processing, and collision delays) and uses this estimate together with location information to calculate the progress speed to each neighboring node. The progress speed is defined as the progress distance (difference between the distances of the current node to destination and next hop to destination) divided by the delay estimate. The neighbor with the greatest progress speed will be chosen as the next hop. As long as the progress speed at every hop is greater than SetSpeed, the end-to-end delay across the network is bound by SetSpeed and the distance between the source and the destination.

The authors of [32] also define a scheduler at the MAC layer of every node that ensures that the delay threshold is not violated. Several priority queues are implemented, each with a predefined SetSpeed. The scheduler picks the packet whose threshold is close to being violated. If no forwarding node with progress speed greater than SetSpeed can be found, the scheduler drops some packets from the queue in order to ensure that the maximum delay is not exceeded. This means that reliability is sacrificed for the sake of speed.

On the other hand, MMSPEED supports reliability of packet delivery by allowing multiple paths to deliver packets. Intermediate nodes determine how many paths should be used based on an error metric. This error metric reflects the probability of packet loss at every node. As the required reliability increases, more paths are utilized to ensure low packet loss rate. However, utilizing multiple paths can increase interference in the network and increase the chance of collisions.

4.3.4.2 Cost and Collision Minimizing Routing

Another cross-layered scheme between network, MAC, and PHY layers was presented in [33], where *cost and collision minimizing routing* (CCMR) was proposed. CCMR addresses the issue of energy waste due to collisions caused by contention at the MAC layer. Routing in CCMR is a hop-by-hop on-demand scheme, where every node is responsible for choosing the next hop that optimizes a certain metric. In order to choose the next hop, a two-dimensional cost metric is proposed. The two dimensions correspond to the probability of successful contention and any other metric that can be of importance to the network. A combination of several metrics can also be implemented. In [33] this metric was chosen to be geographical advancement (obtained from GPS information at the PHY layer) toward the destination. Thus the goal of CCMR in [33] was to choose the next hop that maximizes both the probability of successful contention and geographical advancement toward the destination. CCMR provides a degree of flexibility in choosing routing metrics.

In order to perform routing, the node wishing to transmit sends a RREQ to all active nodes in its one-hop neighborhood. Each node receiving the RREQ first determines if it provides any geographical advancement toward the destination, using location information from the PHY layer. The protocol then defines a group of slots (contention windows) at the MAC layer that can be used to transmit the RREP. If the node receiving the RREQ decides that it can provide geographical advancement to the destination, it will compute which slot it should use to transmit the RREP based on the aforementioned two-dimensional routing metric. The node with the minimum routing metric will have the smallest contention window and will thus transmit the RREP before other nodes, thus being chosen as the next hop. CCMR was shown to be robust and can save energy in case the network load is high by reducing the chance of collisions.

4.3.4.3 Energy-Efficient Multihop Polling in Heterogeneous WSN

Another protocol that addresses the issue of collisions is *energy-efficient multihop polling* (EEMP) [34], which is a cross-layer design between network and MAC layers. EEMP assumes a heterogeneous and hierarchical network structure, where the network is divided into clusters, with each cluster head controlling a group of sensors. The cluster heads are assumed to have higher energy and processing powers than the regular sensors. In addition, EEMP assumes that every node knows its location from GPS or other means.

In the initial phase of EEMP, the clusters are formed. Cluster heads take turns to broadcast their IDs in a control message. Note that the turns of the cluster heads are predetermined according to the IDs given to them. Each cluster head will wait for a predetermined period of time until the previous cluster has finished its local sensor-discovery procedure. Upon hearing a broadcast message, sensor nodes will send a reply to the cluster head using a contention-based MAC protocol. When all the first-hop sensor nodes have sent their replies, the cluster head will ask each of those sensors in turn to send a broadcast packet to determine the second-hop neighbors. Thus, EEMP allows for multiple hops to exist within a cluster. After a sufficient period of time, this process will be repeated several times at every cluster head until every node in the network has been discovered. Note that if a node receives a broadcast message from several clusters, it will only join the one from which the message was received with the highest power.

Once the clusters are formed, nodes can start sending data packets. This is done using a multihop polling algorithm at the network and MAC layers. Since cluster heads do not know in advance which nodes have packets to send, they will start by polling a subset of nodes that cover all the relaying paths of the cluster. Each node, i, in turn will send a data packet (if it has one) in addition to the packet number of any second-hop nodes that are responsible for forwarding their data. After this phase is complete, the cluster head will have acquired some data packets in addition to information about all the remaining packets to be sent. A schedule is then computed at the cluster head that minimizes the number of turns needed to send all the packets. Nodes in the second hop that are not in contention with other nodes in the first hop may send their packets at the same time. This schedule is contention free and minimizes the number of transmissions needed, thus making it energy efficient. In addition, nodes can be given priorities in case a deadline for delay needs to be achieved.

4.3.4.4 Delay Guaranteed Routing and MAC Protocol

In order to minimize collisions and ensure efficiency of transmissions, *delay guaranteed routing and MAC* (DGRAM) was proposed in [35] and utilizes a

TDMA-based MAC layer. Thus DGRAM is a cross-layered approach between network and MAC layers. In contrast to many other TDMA-based MAC protocols, DGRAM does not require a centralized node, and each node determines its transmission slot on its own. In order to perform this task, DGRAM makes the following assumptions: It assumes that all nodes are synchronized, the coverage area of the WSN is circular with the sink node at the center, all nodes know their relative location to the sink, and all nodes discover the location of all other nodes in the network in an initialization phase. The network topology is divided into radial tiers and blocks and, using location information, each node will determine its corresponding tier and block as shown in Figure 4.9.

After each node discovers which tier and block it belongs to, the next step is to determine the TDMA schedule. In order to do this, time is divided into superframes, subframes, sub-subframes, and slots. The size of each time category depends on the number of nodes in the network (which is known to all nodes). A superframe is the period after which the TDMA schedule repeats itself, and contains several subframes. Each tier is assigned a subframe, where two consecutive tiers cannot transmit using the same subframes to avoid collisions. However, nodes that are two tiers apart can transmit in the same subframe.

Each subframe is then divided into sub-subframes according to the number of blocks. Nodes in adjacent blocks cannot transmit using the same sub-subframe. As with subframes, nodes separated by two blocks can transmit using the same sub-subframe. Within the same sub-subframe each node is assigned its own slot according to its radial distance from the sink node (which is known to every node). The node with the smallest radial distance gets the first slot, and so on. If there is a tie in the radial distance, the angular

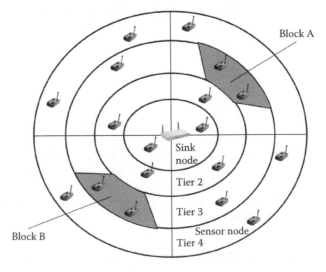

FIGURE 4.9
Tiers and blocks in DGRAM.

distance is used. Therefore, by knowing the locations of nodes and the sink, each node can determine its block and tier, and therefore determine its slot, sub-subframe, and subframe within each superframe. By knowing the exact time when every node can transmit, delay can be accurately bound across the network. In addition, DGRAM is energy efficient since it minimizes collisions.

4.3.5 Limitations to Routing Design across Different Networks

The future of wireless networking is moving toward interconnectivity and compatibility. Wireless users demand seamless connectivity when they move from one network to another. Therefore, it can be expected that QoS support will be required across different networks. For example, a WMN can provide an umbrella of connectivity to cover a large area, such as a university campus, with smaller MANETs covering individual buildings. Research labs may require that their MANET or WMN be connected to WSNs in remote areas, and VANETs can provide exclusive coverage to users moving at high speed between buildings or around the campus.

Connectivity and QoS support in such a heterogeneous network is extremely challenging. Each network type has its own design challenges, which makes it difficult to design a unified framework for QoS support in all networks. However, it is useful to point out some similarities between networks that might enable us to approach the ultimate goal of interoperability.

For example, MANETs and WMNs have several similarities. Both networks suffer from interference challenges and can support challenging applications. However, MANETs demand that all protocols be completely distributed, due to the absence of a centralized node. Therefore, strictly speaking, any distributed protocol can be used for both networks. CACP, AAC, IQRouting, iCAC, and ACA are examples of such protocols. Another solution for interoperability would to be to implement a centralized management system at the WMN that would be responsible for setting up end-to-end routes from source to destination, no matter which network they belong to. The disadvantage of this solution is that it will probably require a gateway to provide compatibility between the two networks. A framework for achieving such compatibility was proposed in [36]. This proposal was based on a generic virtual link layer (GVLL) that would decide on the best serving network to provide QoS support, by considering metrics of throughput, delay, and PLR. However, this solution requires that at least some nodes have a dual interface, one for the WMN and one for the MANET, where the GVLL will hand the incoming packets to the MAC layer of the appropriate interface that would provide the best QoS support.

Although GVLL was proposed to provide connectivity across WMNs and MANETs, its idea can be adapted to connect between WMNs and WSNs as well. However, this issue has multiple challenges due to the strict energy requirements of WSNs. Generally speaking, routing protocols for MANETs, WMNs, and VANETs sacrifice energy to provide QoS support. Thus, simple protocol conversion is not sufficient. Careful consideration of energy has to be included in protocol design. This does not only refer to remaining battery power but also load balancing (dividing the load across multiple nodes) in order to avoid the overuse of a subset of nodes and premature network expiration. If a connection is to be set up between either a WMN or a MANET and a WSN, then different considerations have to be included for the part of the route that belongs to the WMN and MANET, and the part that belongs to the WSN. To our best knowledge, such a task has not yet been attempted in research.

Connectivity between VANETs and other networks also has its challenges. The requirement for high mobility and dynamic topology forces QoS routing design in VANETs to be different from other networks. This essentially means that a protocol that provides interoperability across VANETs and any other network has to be highly adaptive and should respond quickly. In-band signaling is a technique that enables such quick response. Using this technique, control information is piggybacked on data packets to decrease overhead and provide means for constant monitoring of the network state. The INSIGNIA framework [37] is a classic proposal for QoS support using in-band signaling that has been improved and adapted in many ways in order to enhance its performance [38]. Although this proposal was used strictly in MANETs, the idea can be adopted to provide constant monitoring of a VANET in order to adapt to the dynamic topology.

Achieving connectivity between several networks is only part of the problem. End-to-end QoS support across different networks is extremely challenging. First of all, every network is commonly known for requiring certain QoS parameters more than others. For example, delay is the most required parameter in VANETs, PLR is usually the focus in WSNs, while several parameters may be required in MANETS and WMNs. For this reason, research in every network type typically focuses on the most required parameter. However, if we are to consider QoS support and seamless interoperability simultaneously, there has to be research on the support of any parameter in any network. QRDS, RRS, MMSPEED, and CCMR consider more than one metric, but only in the context of a single network. In order to approach the goal of a unified framework for QoS support in any network, research has to start considering the needs of several networks in a generic way and provide means for adapting the protocols as packets move from one network to the other. Such research will ultimately increase as more demands for different networks arise. Already there is significant demand for connectivity between WMNs and MANETs, and this demand can only be expected to increase in the future.

4.4 Comparison between QoS Routing Protocols

In this section we present a comparison between the routing protocols surveyed in this chapter. We compare between the cross-layer approaches taken by the protocols and the aspects of QoS support addressed by each of them, such as the QoS metrics they support and the network constraints they consider. The objective of this section is to provide the reader with an insight into the advantages and disadvantages of the design aspects of every protocol, thus providing a base for future designers of QoS routing protocols for different types of multihop networks. Table 4.2 summarizes some of the aspects of the surveyed protocols.

As Table 4.2 shows, all the surveyed protocols employ aspects of cross-layer design. CACP, RAC, AAC, IQRouting, iCAC, MMSPEED, CCMR, QRDS, and RRS utilize extensive cross-layer interactions between network, MAC, and PHY layers. For example, CCMR calculates the routing metric based on the probability of contention calculated at the MAC layer and GPS information from the PHY layer. The routing protocol then instructs the MAC layer on the size of the contention window that should be used. In QRDS and RRS, extensive PHY layer information is passed to the network layer, which then produces a transmission schedule to be used at the MAC layer. On the other hand, the remaining protocols only utilize PHY or MAC layer information in routing decisions. In addition, the level of cross-layer interaction differs from one protocol to another. For example, the specified VANET protocols only utilize information exchange between network and PHY layers, while a protocol such as EEMP extensively combines the network layer with the MAC layer in order to perform multihop polling. Similarly, all the specified MANET protocols combine network and MAC layers in order to perform routing and call admission, while utilizing information exchange with the PHY layer.

It is interesting to observe that most of the above protocols incorporate PHY layer metrics in routing design. This leads us to believe that PHY layer parameters can have a profound impact on routing and network performance in general. PHY layer parameters are of particular importance in QoS support since several applications require the guarantee of PHY layer resources such as bandwidth or end-to-end delay. Therefore, utilizing realistic PHY layer and propagation models during the design and evaluation stages of routing protocols can be of particular importance. Unfortunately, most of the above protocols (except QRDS, RRS, DGRAM, and EEMP) were evaluated using the disk propagation model, where transmissions are assumed to reach a fixed distance in all directions equally. This model is unrealistic and can lead to inaccurate results.

All the above protocols target QoS support as their main objective. CACP, IQRouting, AAC, iCAC, QRDS, RRS, and ACA all target the QoS metric of throughput. This is because traffic demand in MANETs and WMNs is high and bandwidth is considered a scarce resource. Thus, most protocols proposed for MANETs and WMNs aim for better utilization of the shared medium. Due to the presence of a centralized node in WMNs, QRDS and RRS protocols were designed to exploit this feature. Parameters can be optimized at the centralized node by collecting information from the network, which achieves better performance at the expense of higher complexity. All

TABLE 4.2

Comparison of QoS Routing Protocols

Name	QoS Metric	Disk Propagation Model	Network Targeted/ Requirements	Centralized/ Distributed	Mobility Support	Path Maintenance	Cross-Layer Approach			
							Layers Involved	PHY Parameters	MAC Parameters	
CACP [13]	Throughput	Yes	MANET/ Best QoS Support	Distributed	Limited	No	Network + MAC + PHY	Available Bandwidth + Power Adaptation (CACP-Power)	Call Admission + Contention Count	
RAC [14]	Throughput	Yes	MANET/ Best QoS Support	Distributed	Limited	No	Network + MAC + PHY	Available Bandwidth + Rate Adaptation	Call Admission + Contention Count	
AAC [15]	Throughput	Yes	MANET/ Best QoS Support	Distributed	Yes	No	Network + MAC + PHY	Available Bandwidth	Call Admission + Congestion Level	
IQRouting [16]	Throughput	Yes	MANET/ Best QoS Support	Distributed	No	No	Network + MAC + PHY	Available Bandwidth + Clique Constraints	Call Admission	
iCAC [17]	Throughput + Fairness	Yes	MANET/ Best QoS Support	Distributed	No	No	Network + MAC + PHY	Available Bandwidth	Call Admission	

(continued)

TABLE 4.2 (continued)

Comparison of QoS Routing Protocols

Name	QoS Metric	Disk Propagation Model	Network Targeted/ Requirements	Centralized/ Distributed	Mobility Support	Path Maintenance	Cross-Layer Approach		
							Layers Involved	PHY Parameters	MAC Parameters
QRDS [18]	Throughput + Delay + PLR	No	WMN/ Best QoS Support	Centralized	No	No	Network + MAC + PHY	Interference + Delay + PLR	Reservation Margin + Priority Scheduling
RRS [19]	Throughput + Delay	No	WMN/ Best QoS Support	Centralized	No	No	Network + MAC + PHY	Interference + Traffic Information	Congestion Levels + Scheduling
ACA [20]	Throughput	Yes	WMN/ Best QoS Support	Distributed	No	No	Network + MAC	N/A	Call Admission+ Busyness Ratio + Scheduling
MURU [23]	Link Lifetime + Delay	Yes	VANET/ High Node Mobility	Distributed	Yes	Yes	Network + PHY	GPS Information + Delay	N/A
GVGrid [24]	Link Lifetime	Yes	VANET/ High Node Mobility	Distributed	Yes	Yes	Network + PHY	GPS Information	N/A
GyTAR [25]	Link Lifetime + Throughput	Yes	VANET/ High Node Mobility	Distributed	Yes	Yes	Network + PHY	GPS Information	N/A

(continued)

TABLE 4.2 (continued)
Comparison of QoS Routing Protocols

Name	QoS Metric	Disk Propagation Model	Network Targeted/ Requirements	Centralized/ Distributed	Mobility Support	Path Maintenance	Cross-Layer Approach			
							Layers Involved	PHY Parameters	MAC Parameters
SGPR [26]	Link Lifetime	Yes	VANET/ High Node Mobility	Distributed	Yes	Yes	Network + PHY	GPS Information	N/A
MMSPEED [32]	Delivery Ratio + Delay	Yes	WSN/ Energy-Sensitivity	Distributed	Limited	No	Network + MAC + PHY	Location Information + Delay	Scheduling + Link Reliability
CCMR [33]	Successful Contention + any Cost	Yes	WSN/ Energy-Sensitivity	Distributed	Unspecified	No	Network + MAC + PHY	Location Information	Size of Contention Window
EEMP [34]	Energy Efficiency + Successful Contention	No	WSN/ Energy-Sensitivity	Semi-centralized (clusters)	Limited	No	Network + MAC	N/A	Polling + Scheduling
DGRAM [35]	Energy Efficiency + Successful Contention	No	WSN/ Energy-Sensitivity	Distributed	Limited	No	Network + MAC	N/A	Scheduling

other surveyed protocols utilize distributed operation, which is a requirement due to the nature of MANETs, WSNs, and VANETs. EEMP assumes the presence of sophisticated cluster heads capable of performing centralized operations.

The idea of using the link conflict graph, as in IQRouting, can be accurate in determining the amount of interference present at every link but can be very costly in terms of overhead, especially if the network is large or highly dynamic. A significant advantage of QRDS is that it simultaneously considers several metrics—namely, throughput, delay, and PLR. However, the disadvantage of QRDS is that it was not efficiently analyzed against other QoS protocols. RRS provides an amendment to the protocol that considers delay as well as throughput. ACA only considers the throughput metric but service differentiation between real-time and non-real-time traffic is employed as well. Real-time traffic is prioritized in ACA in order to guarantee a satisfactory QoS level.

On the other hand, MURU considers the QoS metric of delay, which is an important metric for VANET applications such as collision warnings. In addition, MURU, GvGrid, GyTAR, and SGPR consider link lifetime in routing decisions, which is dictated by the nature of VANETs. However, only MURU and GyTAR mention how any of the classical QoS metrics (such as throughput or delay) will be guaranteed in VANETs. Due to network restrictions, all VANET protocols employ path maintenance algorithms to repair broken routes, due to frequent link failures.

Different restrictions are imposed by the nature of WSNs, mainly energy sensitivity and limited capabilities of nodes. To address these issues, the protocols offered are of low complexity. They avoid heavy operations such as optimization. The advantage of MMSPEED is that it considers two QoS metrics: delay and reliability of packet delivery. The disadvantage of MMSPEED and CCMR is that they do not study energy efficiency of the protocols. CCMR provides means of including remaining energy of nodes within the routing metric but does not study this issue when evaluating the performance of the protocol. However, both protocols offer interesting ideas for QoS support in WSNs. EEMP and DGRAM target energy efficiency as their main goal. They propose protocols that minimize contention among nodes, thus ensuring efficient transmissions. In addition, using polling in EEMP and TDMA scheduling in DGRAM provide means of controlling delay of packet delivery in WSNs.

4.5 Challenges and Future Directions

Despite the evolution of protocols targeting QoS support, user applications keep evolving as well, often at a faster pace. Thus, continuous research is needed to accommodate such challenging applications and improve QoS support for the increasing number of wireless users. To reach this goal, some challenges need to be addressed and new directions need to be explored.

Some of the significant challenges that need further investigation include the support of different classes of traffic and the support of Internet services over multihop networks. Although research exists on the support of almost all QoS parameters individually, limited research has dealt with the interactions of nodes when they request the support of different parameters. This is an issue that needs further investigation, especially in MANETs and WMNs, since it is very likely that different applications will be simultaneously requested at different nodes. Some models that have been implemented in the Internet, particularly the differentiated services (DiffServ) model, have been adapted in multihop networks to support QoS. DiffServ is simply a model that supports different levels of services to users, without requiring reservations and without the need for centralized control. It is mainly based on methods that allow intermediate routers to prioritize packets on different flows. A proposal to adapt DiffServ in multihop networks was proposed in [39] by building a backbone that is responsible for transporting high-priority packets.

PHY layer challenges are of particular importance due to the profound impact they have on network performance. Estimating resources such as bandwidth and delay directly impact the efficiency of QoS support. If protocols are too conservative in estimating resources, then those resources may be underutilized. On the other hand, underestimation of resources may cause significant packet loss and degraded QoS levels. Several methods have been proposed to estimate bandwidth. Measuring the channel busy time and SINR are among the most popular methods of measuring bandwidth. Due to the importance of estimating bandwidth, continuous research is needed to improve its accuracy. The amount of research devoted to estimating and supporting end-to-end delay is much less than bandwidth and thus needs to be improved. The main challenge in resource estimation is that resources are shared among interfering nodes that may be outside each other's transmission range, combined with the fact that there is no centralized node capable of monitoring network resources. More research is needed to develop accurate methods of distributed resource estimation.

Another important PHY layer issue that can significantly improve QoS support is the optimization of power and data rate parameters over the entire network, and not just over individual links, in a cross-layer framework. Power control can save energy, optimize network connectivity, and control the level of interference in the network. Data rate control can render links usable if they were not so before, since lower rate transmissions need lower power thresholds than higher rate transmissions and can thus tolerate worse channel conditions.

Exploiting multiple antennas in routing design is also a prominent research direction. Multiple-input/multiple-output (MIMO) is a PHY-layer technology of immense potential as it could lead to significant improvements in capacity and quality of links. Cross-layer design between routing and MIMO has been explored in [40] by controlling the number of antennas used for transmission in order to limit interference and enhance the quality of signals. In [41], a combination of MIMO and adaptive modulation was exploited at the MAC layer to optimize the use of wireless resources or extend the range of signals. Since MIMO enhances signal to noise ratio (SNR), the gain in SNR can either be used to transmit to a longer range or at a higher data rate. There are many ways in which MIMO can be exploited at the network layer, and this idea definitely deserves more attention.

There are also challenges that are often considered secondary (compared to the direct support of QoS metrics) but may have significant impact on QoS support. Challenges such as congestion and fairness are often considered separately, outside of the context of QoS support. Congestion could mean that high-priority traffic is dropped when contending with low-priority traffic. Not considering fairness between nodes could mean that some users may be denied service for the purpose of ensuring high aggregate throughput. TDMA scheduling can be efficient in ensuring fairness and dealing with high traffic loads that may cause congestion but is sometimes difficult to implement in distributed networks, due to their dynamic nature.

A promising direction that is currently being explored by researchers is the introduction of nodes with different capabilities within the same network. This can be of great potential because different wireless standards have different areas of strength and therefore can complement each other, leading to better network performance or a wider range of applications. This idea has been explored in the area of health care in emergency response and disaster scenarios, where a large number of victims can overwhelm medical care providers. The AID-N organization [42] together with the CodeBlue [43] research team at Harvard University proposed a platform for a heterogeneous network that can autonomously monitor vital signs of patients in emergency situations. This network is composed of wireless sensors, ad hoc nodes, and a portal for Internet services. Routing in such a network is complicated and deserves more attention. The capabilities and limitations of the different node types must be taken into consideration in order to realize the full potential of the network. Heterogeneous nodes can also be used in WSNs, for example, to alleviate congestion or save energy by using more powerful nodes to transmit to longer distances or to process data at intermediate nodes.

Supporting QoS is particularly challenging in WSNs due to the limited energy and processing resources. Therefore research on QoS routing in WSNs will be ultimately hindered by hardware limitations. Energy harvesting and renewable energy resources (such as solar cells) [44] are highly important to address limited energy resources. As research on energy-efficient hardware improves, QoS support may improve as well. However, until hardware improves, one of the most promising directions for improving network performance is employing cognitive approaches in multihop networks [45]. Cross-layered approaches suffer from drawbacks such as adaptation loops, lack of adaptability to changing network conditions, slow reaction, and lack of capabilities to support multiple conflicting objectives.

As highlighted in this book, cognitive approaches can make a breakthrough in this area as they enable intelligence at different layers in the network in order to improve adaptability and achieve network's end-to-end goals. At the PHY layer, cognitive radios can alleviate the problem of interference and provide a boost in the availability of wireless resources by enabling opportunistic and dynamic use of the radio spectrum. This can be of particular importance in MANETs and WMNs, where traffic is expected to be heavy. At upper layers, a cognitive engine can provide means for nodes to learn and adapt their parameters according to the application requirements and the dynamic wireless environment. This will certainly open new doors for QoS support in challenging networks such as WSNs and VANETs, and help achieve network goals in the presence of multiple conflicting objectives.

4.6 Conclusions

In this chapter a comprehensive study of QoS routing design in multihop wireless networks was presented. Four types of multihop networks—namely, MANETs, WMNs, VANETs, and WSNs—were considered, and focus was given to cross-layered protocols due to their efficiency. The main challenges in routing design of every type of multihop network were highlighted, and protocols that address these challenges were studied.

Several problems still have not been fully addressed in multihop networks. Such problems include service differentiation, accurate resource estimation, realistic PHY layers and propagation models, and power and data rate control. Cognitive communications also represent a new direction that promises to deliver solutions to several persistent networking problems, such as crowded spectrums and conflicting network goals. QoS support is certainly a topic of significant importance and will probably continue to receive great attention from the research community in the future.

References

1. S. Basagni, M. Conti, S. Giordano, and I. Stojmenovic, *Mobile Ad Hoc Networking*, John Wiley and Sons, 2004.
2. R. Shorey, A. Ananda, M.C. Chan, and W.T. Ooi, *Mobile, Wireless and Sensor Networks, Technology, Applications and Future Directions*, John Wiley and Sons, 2006.

3. I. Stojmenovic, *Handbook of Sensor Networks, Algorithms and Architectures,* John Wiley and Sons, 2005.
4. Q. Zhang and Y.-Q. Zhang, "Cross-layer design for QoS support in multihop wireless networks," *Proceedings of the IEEE* 96, no. 1, January 2008, pp. 64–76.
5. L. Chen and W. Heinzelman, "A survey of routing protocols that support QoS in mobile ad hoc networks," *IEEE Network Journal* 21, no. 6, November 2007, pp. 30–38.
6. L. Hanzo II and R. Tafazolli, "Admission control schemes for 802.11-based multi-hop mobile ad hoc networks: A survey," *IEEE Communications Surveys and Tutorials*, 11, no. 4, December 2009, pp. 78–108.
7. F. Li and Y. Wang, "Routing in vehicular ad hoc networks: A survey," *IEEE Vehicular Technology Magazine* 2, no. 2, June 2007, pp. 12–22.
8. C. Perkins and E. Royer, "Ad-hoc on-demand distance vector routing," *Second IEEE Workshop on Mobile Computing Systems and Applications* (WMCSA), February 1999, pp. 90–100.
9. C. Perkins and P. Bhagwat, "Highly dynamic destination-sequenced distance vector (DSDV) routing for mobile computers," *Proceedings of the Conference on Communications Architectures, Protocols, and Applications (SIGCOMM)* 24, no. 4, October 1994, pp. 234–244.
10. D. De Couto, D. Aguayo, J. Bicket, and R. Morris, "A high throughput path metric for multi-hop wireless routing," *Proceedings of the 9th International Conference on Mobile Computing and Networking (MobiCom),* September 2003, pp. 134–146.
11. L. Ma and M. Denko, "A routing metric for load balancing in wireless mesh networks," *21st IEEE International Conference on Advanced Information Networking and Applications (AINAW)* 2, May 2007, pp. 409–414.
12. A.J. McAuley, K. Manousakis, and L. Kant, "Flexible QoS route selection with diverse objectives and constraints," *16th IEEE International Workshop on Quality of Service (IWQoS),* June 2008, pp. 279–288.
13. Y. Yang and R. Kravets, "Contention-aware admission control for ad hoc networks," *IEEE Transactions on Mobile Computing* 4, no. 4, July 2005, pp. 363–377.
14. L. Luo, M. Gruteser, H. Liu, D. Raychaudhuri, K. Huang, and S. Chen, "A QoS routing and admission control scheme for 802.11 ad hoc networks," *International Conference on Mobile Computing and Networking,* September 2006, pp. 19–28.
15. R. de Renesse, R. Friderikos, and H. Aghvami, "Cross-layer cooperation for accurate admission control decisions in mobile ad hoc networks," *IET Communications* 1, no. 4, August 2007, pp. 577–586.
16. R. Gupta, Z. Jia, T. Tung, and J. Warland, "Interference-aware QoS routing (IQRouting) for ad hoc networks," *IEEE Global Telecommunications Conference (GLOBECOM),* November 2005, pp. 2599–2604.
17. K. Sridhar and M. Chan, "Interference based call admission control for wireless ad hoc networks," *IEEE International Conference on Mobile and Ubiquitous Systems,* July 2006, pp. 1–10.
18. C.-H. Liu, A. Gkelias, and K.-K. Leung, "A cross-layer framework of QoS routing and distributed scheduling for mesh networks," *IEEE Vehicular Technology Conference (VTC),* May 2008, pp. 2193–2197.
19. W. Wang, X. Liu, and D. Krishnaswamy, "Robust routing and scheduling in wireless mesh networks under dynamic traffic conditions," *IEEE Transactions on Mobile Computing* 8, no. 12, December 2009, pp. 1705–1717.

20. Q. Shen, X. Fang, P. Li, and Y. Fang, "Admission control based on available bandwidth estimation for wireless mesh networks," *IEEE Transactions on Vehicular Technology* 58, no. 5, June 2009, pp. 2519–2528.
21. P. Pathak and R. Dutta, "A survey of network design problems and joint design approaches in wireless mesh networks," *IEEE Communications Surveys and Tutorials,* 2010, pp. 1–33.
22. M. Boban, G. Misek, and O. Tonguz, "What is the best achievable QoS for unicast routing in VANETs?" *IEEE Global Telecommunications (GLOBECOM) Workshops,* November 2008, pp. 1–10.
23. Z. Mo, H. Zhu, K. Makki, and N. Pissinou, "MURU: A multi-hop routing protocol for urban vehicular ad hoc networks," *3rd IEEE Annual International Conference on Mobile and Ubiquitous Systems: Networking and Services,* July 2006, pp. 1–8.
24. W. Sun, H. Yamaguchi, K. Yukimasa, and S. Kusumuto, "GVGrid: A QoS routing protocol for vehicular ad hoc networks," *14th IEEE International Workshop on Quality of Service (IWQoS),* June 2006, pp. 130–139.
25. M. Jerbi, S. Senouci, T. Rasheed, and Y. Ghamri-Doudane, "Towards efficient geographic routing in urban vehicular networks," *IEEE Transactions on Vehicular Technology* 58, no. 9, November 2009, pp. 5048–5059.
26. T. Taleb, E. Sakhaee, A. Jamalipour, K. Hashimoto, N. Kato, and Y. Nemoto, "A stable routing protocol to support ITS services in VANET networks," *IEEE Transactions on Vehicular Technology* 56, no. 6, November 2007, pp. 3337–3347.
27. D. Shi, X. Zhang, X. Gao, W. Zhu, and F. Zou, "A link reliability-aware route maintenance mechanism for mobile ad hoc networks," *IEEE International Conference on Networking (ICN),* April 2007, pp. 8–14.
28. C.-Y. Chang and S.-C. Tu, "Active route-maintenance protocol for signal-based communication path in ad hoc networks," *Journal of Network and Computer Applications* 25, no. 3, 2002, pp. 161–177.
29. Y.-H. Wang, C.-C. Chuang, C.-P. Hsu, and C. Chung, "Ad hoc on-demand routing protocol setup with backup routes," *IEEE International Conference on Information Technology: Research and Education,* August 2003, pp. 137–141.
30. W. Heinzelman, A. Chandrakasan, and H. Balakrishnan, "Energy-efficient communication protocol for wireless microsensor networks," *Proceedings of the 33rd Hawaii International Conference on System Sciences (HICSS)* 8, January 2000, pp. 1–10.
31. M. Zorzi and R. Rao, "Geographic random forwarding (GeRaF) for ad hoc and sensor networks: Multihop performance," *IEEE Transactions on Mobile Computing* 2, no. 4, October 2003, pp. 337–348.
32. E. Felemban, C.-G. Lee, and E. Ekici, "MMSPEED: Multipath multi-SPEED protocol for QoS guarantee of reliability and timeliness in wireless sensor networks," *IEEE Transactions on Mobile Computing* 5, no. 6, June 2006, pp. 738–754.
33. M. Rossi, N. Bui, and M. Zorzi, "Cost and collision minimizing forwarding schemes for wireless sensor networks: Design, analysis, and experimental validation," *IEEE Transactions on Mobile Computing* 8, no. 3, March 2009, pp. 322–337.
34. Z. Zhang, M. Ma, and Y. Yang, "Energy efficient multihop polling in clusters of two-layered heterogeneous sensor networks," *IEEE Transactions on Computers* 57, no. 2, February 2008, pp. 231–245.
35. C. Shanti and A. Sahoo, "DGRAM: A delay guaranteed routing and MAC protocol for wireless sensor networks," *IEEE Transactions on Mobile Computing* 9, no. 10, October 2010, pp. 1407–1423.

36. J. Roy, V. Vaidehi, and S. Srikanth, "Always best-connected QoS integration model for the WLAN, WiMAX heterogeneous network," *IEEE International Conference on Industrial and Information Systems,* August 2006, pp. 361–366.

37. S.-B. Lee and A. Campbell, "INSIGNIA: In-band signaling support for QoS in mobile ad hoc networks," *International Workshop on Mobile Multimedia Communication (MoMuc),* October 1998, pp. 1–12.

38. Y. He and H. Abdel-Wahab, "HQMM: A hybrid QoS model for mobile ad hoc networks," *IEEE Symposium on Computers and Communications,* June 2006, pp. 194–200.

39. M. Fazio, M. Paone, D. Bruneo, and A. Puliafito, "Cross-layer architecture for differentiated service in ad hoc networks," *IEEE International Symposium on Network Computing and Applications,* July 2008, pp. 128–135.

40. A. Gkelias, F. Boccardi, C. Liu, and K. Leung, "MIMO routing with QoS provisioning," *IEEE International Symposium on Wireless Pervasive Computing (ISPWC),* May 2008, pp. 46–50.

41. E. Gelal, G. Jakllari, and S. Krishnamurthy, "Exploiting diversity gain in MIMO equipped ad hoc networks," *40th Asilomar Conference on Signals, Systems and Computers,* October 2006, pp. 117–121.

42. T. Gao, T. Massey, L. Selavo, D. Crawford, B.-R. Chen, K. Lorincz, V. Shnayder, L. Haunstein, F. Dabiri, J. Jeng, A. Chanmugam, D. White, M. Sarrafzadeh, and M. Welsh, "The advanced health and disaster aid network: A light weight wireless medical system for triage," *IEEE Transactions on Biomedical Circuits and Systems* 1, no. 3, September 2007, pp. 203–216.

43. T. Gao, C. Pesto, L. Selavo, Y. Chen, J. Ko, J. Lim, A. Terzis, A. Watt, J. Jeng, B.-R. Chen, K. Lorincz, and M. Welsh, "Wireless medical sensor networks in emergency response: Implementation and pilot results," *IEEE International Conference on Technologies for Homeland Security,* May 2008, pp. 187–192.

44. V. Raghunathan, A. Kansal, J. Hsu, J. Friedman, and M. Srivastava. "Design considerations for solar energy harvesting wireless embedded systems. *4th International Symposium on Information Processing in Sensor Networks,* April 2005, pp. 457–462.

45. G. Vijay, E. Bdira, and M. Ibnkahla, "Cognition in wireless sensor networks: A perspective," *IEEE Sensors Journal* 11, no. 3, 2011, pp. 582–592.

5

Cognitive Diversity Routing

5.1 Overview of Routing Protocols in Wireless Sensor Networks

5.1.1 Wireless Sensor Network Routing Protocols

Sensor nodes are constrained in terms of processing, storing, and most importantly, energy. The most important characteristics in a routing protocol for a sensor network are energy efficiency and awareness. Multihopping is widely used in wireless sensor networks (WSNs) due to the limited transmission range and is also more energy efficient due to signal attenuation. There are mainly three types of routing protocols for wireless sensor networks: flat-based, hierarchical, and location-based (Al-Karaki and Kamal 2004). Flat-based routing is when all nodes have the same functionality. In hierarchical routing, nodes have different roles where the aim is to cluster nodes so that there will be aggregation at the cluster heads in order to reduce energy consumption. Location-based routing aims at exploiting the node position information for routing purposes.

Sensor protocol for information via negotiation, or SPIN, is a flat-based routing protocol based on negotiations and resource adaptation (Al-Karaki and Kamal 2004). These two notions were introduced in order to combat the faults of the flooding approach. SPIN enables nodes to negotiate with each other before transmitting data to ensure that only useful information is transmitted in the network, thus considerably saving energy consumption. In order to negotiate, nodes use metadata and high-level descriptors for describing the data they want to transmit. Metadata describes the actual data being collected by the sensor. Therefore there is no standard metadata format since it will vary from one application to another. The size of the metadata is much smaller than that of a data packet; otherwise it would defeat the objective of the protocol. SPIN first sends an advertising message about the new data, then neighbors that want to obtain the data send a request data message to which the node replies with the actual data. This way, the actual data are only sent to nodes that need the data, thus significantly decreasing the number of transmitted messages, which means considerable amounts of energy

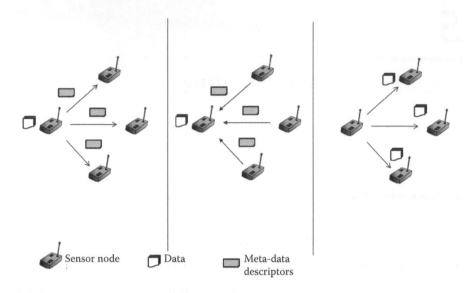

FIGURE 5.1
SPIN protocol negotiation process.

saved (see Figure 5.1). The main disadvantage for SPIN is that the negotiation process may result in distant nodes unable to obtain the data if intermediate nodes do not require it. It is also not a suitable protocol for applications where data packets are required to be delivered over regular intervals.

Directed diffusion protocol is a data-centric protocol in which the sink issues a query to a certain region so the nodes in that region cooperate and gather the data between themselves to send the data back to the sink (Verdone et al. 2008). One of the main aspects of this protocol is utilizing a naming scheme for data gathered by the sensors in order to save energy while eliminating some unnecessary operations in the network layer. The mode of operation of this protocol consists of the sink first issuing a sensing task that is disseminated over the network in a form of an interest for named data. Nodes matching the interest will respond and data are sent to the sink through multiple paths where intermediate nodes cache and aggregate data. The interests are refreshed and updated on a periodic basis. The drawbacks of this protocol are that it does not work for applications that require continuous data transmissions to the sink since the protocol is query based. Naming schemes are application specific so they have to be customized each time. Moreover, the matching process for data and queries requires extra overhead.

Power-efficient gathering in sensor information system (PEGASIS) is a hierarchical WSN routing protocol (Al-Karaki and Kamal 2004). PEGASIS forms a chain of sensor nodes instead of having multiple clusters and cluster heads (see Figure 5.2). Each node will then transmit to the nearest node in the chain and then only one node in the chain will be designated to transmit to the sink. Nodes determine their closest neighbors by measuring signal strengths

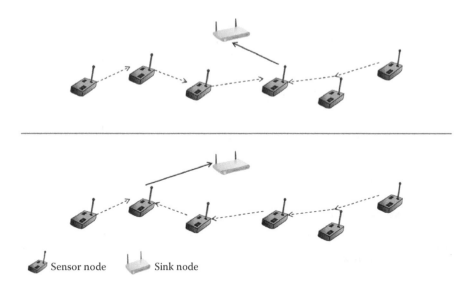

FIGURE 5.2
PEGASIS chain of nodes.

and then the neighbors change their own signal strength so that it only reaches the nearest neighbor. Each node receives the data from its neighbor, aggregates it with its own data, and then forwards the data to the next node in the chain. The chain will be reconstructed whenever a node dies. In some random deployments, neighboring nodes will not be neighbors in the chain and that results in them using more energy for transmission. This protocol was proposed as an improvement to the LEACH protocol that forms clusters whereby cluster heads transmit the data collected from their cluster members to the sink. The main disadvantages lie with the excessive delay for distant nodes in the chain. In addition, having only one designated leader increases the chance of bottlenecks. Also, this protocol requires dynamic topology adjustments because every node is only aware of its neighbors.

Geographic random forwarding (GeRaF) is a location-based routing WSN protocol that forwards data based on geographical location of the nodes and random selection of the relaying node by having contention among the receivers (Verdone et al. 2008). Each node knows its position and the position of the sink (assuming, for example, that they are equipped with GPS receivers). When a node wants to transmit to the sink, it sends a broadcast specifying its location and that of the sink (see Figure 5.3). Then the neighboring nodes, upon hearing the broadcast, set their next hop priorities based on their distance to the destination. The relays continue in this process until data reach their destination. The neighboring nodes determine their own distance to the sink in order to determine their suitability as relay nodes. The suitability is determined by dividing the coverage area into two regions: the relay region and the nonrelay region. The relay region consists of the area within

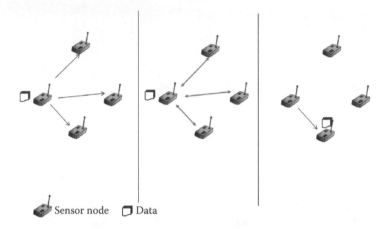

Sensor node ☐ Data

FIGURE 5.3
Geographic random forwarding.

closer range to the destination than the transmitting node, while the nonrelay region contains the other part. A node can only be selected as a relay if it is in the relay region. The relay region is further divided into priority regions based on the distance from the destination. The main disadvantage is the lack of consideration for any factor other than distance and that may lead to bottleneck situations. There is also a distinct lack of reliability mechanism in the protocol that guarantees delivery.

Geographic adaptive fidelity (GAF) is an energy-aware location-based routing protocol that can be used for mobile ad hoc networks as well as wireless sensor networks (Verdone et al. 2008). GAF divides the network into equal square zones, forming a virtual grid. Each node is then associated with a zone in the grid based on its position (for example, determined by its GPS receiver). GAF operates by keeping the least amount of nodes required in the zone awake and turning all the others in that zone to the sleeping mode. The nodes in the zone coordinate and decide on the length and the period of the sleeping cycle. The node that is awake, or the zone leader, collects the data for the nodes in its zone and is responsible for gathering and transmitting data to the sink by routing through other awake nodes. The designated node is very similar to the leader in the hierarchical routing protocol, so GAF can be considered as a hybrid protocol that is both location based and hierarchical. Nodes alternate leaders in order to balance the overall energy consumption. The main drawback of this protocol is having only the grid leader transmitting, which may lead to bottlenecks, congestion, and longer distances for the transmitting nodes.

5.1.2 Energy-Aware Protocols

Energy-aware protocols take into account the signal strength to reduce energy consumption by choosing optimal forwarding nodes.

For example, Frey, Ruhrup, and Stojmenovic (2009) proposed a power-aware routing protocol that defined a general power metric that consists of the signal attenuation, startup energy loss, retransmissions, and collisions in a single expression depending on the distance between the sender and receiver. The protocol assumes that the power consumption for a path is equal to the optimal one; each intermediate node will select a neighbor closer to the destination in order to minimize the sum of power needed to transmit the packet and the optimal power consumption needed to forward the packet from the transmitting node to the destination. Therefore, an optimal number of intermediate forwarding nodes produce minimal power consumption, and that optimal number is found from the distance between the two nodes and general power metric parameters. Power routing attempts to minimize the energy consumption, but a node can be chosen for several routing paths, which may result in heavy utilization at that node and may cause its death. The cost metric used in this protocol is inversely proportional to the remaining battery power; the forwarding node seeks to minimize the sum of the cost metric and the estimated cost for the remaining path.

The localized power and cost-aware routing scheme proposed in Kuruvila, Nayak, and Stojmenovic (2006) is based on the idea of proportional progress. In this protocol, the neighbor that minimizes an expression containing the signal attenuation exponent and the minimal reception and computational power is selected to forward the message, thereby minimizing the power spent in every unit of progress made. The node that has the packet will then forward it to a neighbor closer to the destination, which results in minimizing the ratio of power consumption and a cost metric (obtained by a proposed algorithm) to reach that neighbor. The progress made is measured as a reduction or projection along the line to the destination. The cost metric is a function of the distance d between the transmitter and destination, r between transmitter and neighbor, x between neighbor and destination, and $f(A)$, which is the reluctance of the transmitting node to forward a packet. The function that the transmitting node seeks to minimize is $f(A) / (d - x)$ because the minimum result yields the next hop that has the least cost.

Geographic and energy-aware routing (GEAR) is a routing protocol that combines location-based routing with data-centric routing (Verdone et al. 2008). This protocol is heavily used by location-aware networks where they need to send specific information to a specific geographic location (see Figure 5.4) GEAR is expected to be more energy efficient than other data-centric protocols in the sense that it only directs the queries to the desired locations instead of broadcasting them over the network. This protocol is only directed at applications where data are gathered upon request rather than periodically sent to the sink. GEAR minimizes the number of interests presented in the directed diffusion protocol, sending the queries directly to the area of interest instead of broadcasting the interest in the network. GEAR also utilizes energy-aware neighbor selection when routing toward the destination and uses recursive geographic forwarding to send the interest packets inside the desired region.

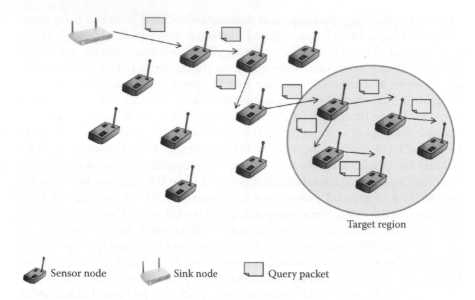

Target region

Sensor node   Sink node   Query packet

FIGURE 5.4
GEAR protocol: Query packet forwarding to target region.

The *dynamic election-based sensing and routing approach* presented in Oteafy, AboElFotoh, and Hassanein (2009) is a novel approach in sensing in WSNs where a single node gets elected to report an event and forward data (see Figure 5.5). This process is performed in a decentralized manner by having the neighboring node that is the fit elected to avoid redundancy and unnecessary transmissions. In this protocol, fitness is the reliability of the node, which is the probability of the message successfully reaching the sink. Node fitness is determined by its hop value and battery energy whereby each parameter is assigned a weight depending on its importance. If a heavier weight is assigned to the battery value, load distribution over the network occurs, while if the hop value is given a higher weight, the node will undertake the shortest path to the destination (in terms of number of hops). After a node determines its fitness, it sets a timer that is inversely proportional to its fitness, after which it declares itself as the most suitable node to forward the packet and the process is repeated until the destination is reached.

The *energy-aware routing protocol* proposed in Shah and Rabaey (2002) is aimed to extend the network's lifetime. This protocol is similar to the directed diffusion protocol presented earlier; however, it differs by having several paths instead of one optimal path. The paths are chosen by the mean of a certain probability that depends on the level of energy consumption in a path. The protocol is composed of three phases. The first is the setup phase, where all nodes find routes to the destination as well as their respective energy costs and construct tables accordingly. The next phase is the data communication phase, where paths are chosen probabilistically based on the

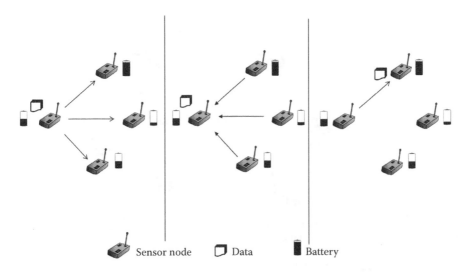

Sensor node　　　Data　　　Battery

FIGURE 5.5
Dynamic election-based routing.

energy costs in the first phase. Finally, the third phase consists of the route maintenance, where localized flooding is performed sporadically to keep the routing tables updated.

Local update-based routing protocol (LURP) (Wang et al. 2007) has been formulated around the idea of taking advantage of the sink's mobility. In this protocol, when a sink moves, it only broadcasts its information to a local area instead of the entire network in order to consume less energy and decrease the probability of collision in the network. This protocol is hierarchical and location based, and the way it operates is by having the sink in the virtual center (Wang et al. 2007). When a node wants to send a packet to the sink, it uses a geocasting protocol to a node in the virtual center. Then, this node uses a topology-based protocol to forward the packet to the sink. This protocol significantly reduces the number of broadcast packets in the network, thus decreasing energy consumption and congestion and, at the same time, utilizing the advantages of two of the three types of sensor-routing protocols.

5.1.3 Diversity Routing

Diversity is often used in communication systems to combat fading in wireless channels; hence this technique is employed to increase reliability. Time diversity, frequency diversity, space diversity, and multiuser diversity have all been used to combat fading. However, in Shariatpanahi and Aarabi (2007) another form of diversity is exploited, one that is inherent in any wireless network due to the shared medium of propagation. In a network, when a node transmits to another, other nodes are able to hear the transmission due to the nature of the medium; hence this data can be used to increase the reliability and assist in the

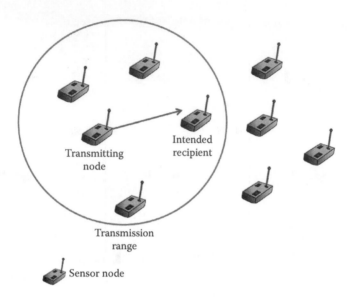

FIGURE 5.6
Diversity routing.

transmission. This notion was also exploited in Lenders and Baumann (2008) where a link-diversity routing paradigm was presented. The link-diversity routing protocol chooses each hop of a route based on the number of outgoing links toward the destination at the intermediate nodes. The decision maximizes the probability of success at each hop in the presence of link failures caused by fading, which yields more reliable paths utilized for routing after avoiding links that are more prone to failure. Diversity routing is also utilized in Mansouri et al. (2005) to reduce the transmission power and prolong network lifetime (see Figure 5.6). In WSNs, path redundancies occur often to ensure reliability and guarantee transmission to the sink node, thereby taking advantage of this diversity to decrease overall power consumption. This is done by establishing a tradeoff between path redundancy and fault tolerance in the network since they are directly related.

Diversity routing is heavily utilized in cooperative routing. Cooperative diversity is obtained through a sequence of sets of cooperating nodes along with appropriate allocation of transmission power that has been shown to mitigate the effect of fading.

5.1.4 Cognitive Protocols

With the increased use of cognitive techniques (Vijay, Bdira, and Ibnkahla 2010; Fortuna and Mohorcic 2009; Akan et al. 2009; Boonma and Suzuki 2007; Niezen et al. 2007; Reznick and Von Pless 2008; Shufhiang 2005) in WSNs, several protocols have been developed to implement cognitive aspects. The *cognitive channel-aware routing* (CCAR) proposed in El Mougy, El Jabi, et al. (2010)

addresses the issue of high packet loss in WSNs. This energy-efficient proto-col attempts to reduce the packet loss ratio, thereby reducing the number of retransmissions. When a node wants to transmit to the sink, it broadcasts an RREQ (adapted to the node density in the network to minimize transmis-sion power). Then, by obtaining the channel profile of all the neighbors, it can determine the minimum power needed to reach a certain neighbor. The path is selected by combining two metrics, namely, the transmission power needed and the packet loss ratio.

In Felemban, Lee, and Ekici (2006), a multipath protocol called multispeed (MMSPEED) is proposed. This protocol utilizes information from the physi-cal and MAC layers to assist the network layer. Reliability, delay, and loca-tion information are relayed to the network layer to determine the packet forwarding. Each node maintains an estimate of the delay incurred to trans-mit to each of its neighbors, uses this information along with the location information to calculate the cost to each neighboring node, and then chooses the optimal next hop. This protocol allows the intermediate nodes to deter-mine the number of paths based on a predefined metric. MMSPEED utilizes localized mechanisms for quality of service (QoS) provisioning by using geographic packet forwarding with compensations for local decision inac-curacies as packets are forwarded to the destination, thereby guaranteeing end-to-end requirements that ensure scalability and adaptability.

Another proposed cognitive protocol is the *cost and collision minimizing routing* (CCMR) (Rossi, Bui, and Zorzi 2009), which also involves the physi-cal, network, and MAC layers. CCMR examines energy losses in collisions caused by MAC layer contentions. CCMR's objective is to determine the next hop protocol that has the minimal cost by maximizing the probability of a successful contention. The cost is based on the metric that determines the node's distance to the destination. In order to minimize collisions, the transmitting node sends a RREQ to determine the path to the destination. The nodes that receive the request then proceed to determine the contention window to transmit the reply in a manner to avoid collision and maintain successful contention. The node with the smallest contention window will transmit the reply before the others, thus reducing the energy losses due to minimized collisions.

In the following, we present cognitive diversity routing that was first pro-posed by El-Jabi (2010).

5.2 System Models

This section presents the models to be adopted for the proposed system. These models specify the environments where the system will be operating. In this section, the energy and propagation models are presented and the network lifetime definition is explored and quantified, since extending the network lifetime is the main objective of this protocol.

5.2.1 The Propagation Model

Propagation models are essential for predicting the path loss along a link and evaluating the surrounding environment and conditions. Obtaining information regarding the quality and state of the link between two wireless nodes can aid in determining the feasibility of the link and how much power will be lost as a result. The two main determining factors of a propagation channel are the path loss and the fading of the link. The path loss is a decibel ratio of transmitted power to received power; it serves greatly in determining the energy consumption of a transmission process. There are two types of statistical path loss models: the deterministic and the statistical. The deterministic model expresses the average received power only in terms of the distance between the transmitting and the receiving node

$$\overline{P}(d) = \overline{P_0}\left(\frac{d}{d_0}\right)^{-\alpha} \qquad \text{(linear scale)}$$

$$\overline{P}_{dB}(d) = \overline{P}_{dB}(d_0) - 10\ \alpha \log\left(\frac{d}{d_0}\right)$$

and the path loss (dB) is expressed as:

$$\overline{PL}_{dB}(d) = \overline{PL}_{dB}(d_0) + 10\ \alpha \log\left(\frac{d}{d_0}\right),$$

where α is the path loss exponent, usually between 2 and 5, depending on the environment, d is the distance from the receiver, d_0 is a reference distance.

The statistical model takes into account the changes in the surrounding environment. This is known as lognormal shadowing, which is characterized by a Gaussian random variable with zero mean and variance, σ^2. The path loss (in dB) is determined as (Rappaport 2001; El Mougy, Bdira, and Ibnkahla 2010):

$$PL_{dB}(d) = \overline{PL}_{dB}(d) + X_\sigma = \overline{PL}_{dB}(d_0) + 10\ \alpha \log\left(\frac{d}{d_0}\right) + X_\sigma,$$

and the received power (dB) is expressed as:

$$P_{dB}(d) = \overline{P}_{dB}(d) + X_\sigma = \overline{P}_{dB}(d_0) - 10\ \alpha \log\left(\frac{d}{d_0}\right) + X_\sigma$$

Several energy models have been derived in the literature from the above propagation model (see, for example, Kim B. & Kim I. 2006; Griva et al. 2009).

5.2.2 Network Lifetime

Nodes in wireless sensor networks are characterized by having a finite and limited energy source, especially in control or monitoring applications. Hence, once they are deployed for operation, it is vital that they last for a rational and practical amount of time; otherwise, it would render the entire technology obsolete. Intensive research is directed into prolonging the period of operation of the nodes and the network as a whole in order to increase the performance of wireless sensor networks. There are two important concepts in this domain: node lifetime and network lifetime. Node lifetime is basically the period of operation of a node. Each node is equipped with an energy source, such as a battery or an alternative source. The node lifetime is also defined by the time during which the energy level of the node is above a certain threshold that allows it to perform its operations, such as transmitting, receiving, or processing data. Once the energy level drops below this threshold, the node is considered dead and is therefore unable to operate.

Network lifetime, on the other hand, is a more complex notion and varies from one application to another. In some applications, every node is vital; hence the failure of one node leads to the failure of the entire network. In other larger applications, there is redundancy in the network, which means that there are several nodes that carry out the same function or monitor the same area. Generally speaking, in most networks it is tolerable to lose certain nodes as long as the backbone of the network is still able to function. There have been many definitions of network lifetime, as it depends on the applications and deployment conditions; however, the most commonly used definitions are:

- *Network Lifetime Definition:* The interval of time, starting with the first transmission in the wireless network and ending when the percentage of alive nodes falls below a specific threshold, which is set according to the type of application (Verdone et al. 2008).

Analyzing the above definition in depth, while taking into consideration that an operational network is composed of active nodes and links, the network lifetime definition can further be broken down into two definitions: the connectivity-based (CB) and percentage of alive nodes (PAN) definitions.

- *Connectivity-Based Definition:* Lifetime of a WSN is the time span from deployment to the instant when a network partition occurs (Al-Turjman, Hassanein, and Ibnkahla 2009). An example of network partition is displayed in Figure 5.7.

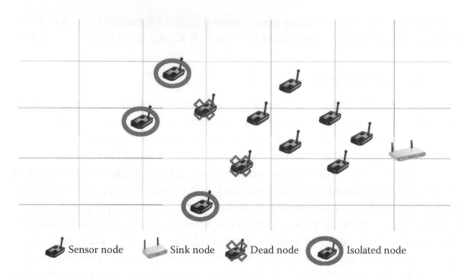

FIGURE 5.7
Network partition.

- *Percentage of Alive Nodes Definition:* Lifetime of a WSN is the time span from deployment to the instant when the percentage of live nodes falls below a specific threshold (Al-Turjman, Hassanein, and Ibnkahla 2009).

After defining the network lifetime, the next step would be to obtain performance metrics in order to be able to quantify the network lifetime definition. The most common metrics used are:

- *First Sensor Node Death (FND):* This metric is a simple approach defined by recording the time when one of the nodes becomes no longer operational. This metric is used in applications where each node is critical, but it is useful in other applications to give a lower bound on the lifetime definition (Ozgovde and Ersoy 2007).
- *Ratio of Remaining Energy (RRE):* Ratio of total energy amount over all the nodes (Al-Turjman, Hassanein, and Ibnkahla 2009).
- *Probability of Node Isolation* (Mizanian, Yousefi, and Jahangir 2009).
- *Probability of Node Failure* (Al-Turjman, Hassanein, and Ibnkahla 2009; Verdone et al. 2008).

These definitions are used to quantify network lifetime in order to carry out performance evaluations on different protocols. This will help evaluate the different protocols and determine the most suitable one for extending the network lifetime.

5.3 Cognitive Diversity Routing

This section presents an energy-efficient routing protocol for disseminating data by selecting routes based on their energy, channel, and traffic states. The protocol was presented first in El-Jabi (2010). Due to the transmission process being responsible for most of the energy consumption, this protocol was designed to minimize the number of transmissions in the network as a whole, thus extending the lifetime of the network—which is the primary target for this protocol. The cognitive diversity routing protocol is, as the name implies, a protocol that combines the two notions of cognitive and diversity routing.

5.3.1 Cognitive Diversity Routing Methodology

Cognitive diversity routing (CDR) is divided into three main phases: the initial phase, the network maintenance phase, and the routing phase. In the initial phase, the basic backbone of the network is formed and the nodes establish communications with each other. The network maintenance phase ensures that the network is kept continuously updated of any changes that happen to the nodes, which is important to prevent information from being outdated. The routing phase is the most important phase in which decisions are made to disseminate data and select routes from the source to the destination. The cognitive diversity routing protocol was chosen to be a proactive routing protocol. Proactive routing protocols are table-driven protocols that maintain updated lists of their destinations and routes by periodically updating the routing tables throughout the network. Figure 5.8 illustrates cognitive diversity routing.

5.3.1.1 Phase I: Initial Discovery

The first phase in the cognitive diversity protocol is the initial discovery where the different nodes in the network establish communication and initialize their routing tables. The first task a node undertakes is to calculate its current energy. The approach where information is obtained from the physical layer is what makes this routing protocol cognitive and intelligent.

The next step would have nodes begin transmitting information packets to all their neighbors in order to exchange information and learn about the other nodes in the network. This is done through flooding, where each node transmits a flooding packet that has the format shown in Figure 5.9.

This format allows for the node to have GPS capabilities so that it can also transmit geographic information in order to assist with its routing process.

After the flooding packets have been transmitted, each node is now capable of constructing routing tables to hold all the information about the nodes in the network. Two tables are constructed: the neighbor table and

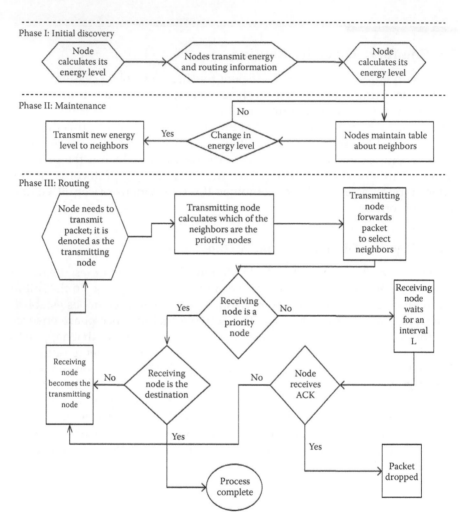

FIGURE 5.8
System model of cognitive diversity routing.

the destination table. The neighbor table is for the node to store information about its neighboring nodes, where it is utilized afterward in the routing process when determining the next hop. The neighbor table incorporates the energy level and the traffic levels of the neighboring nodes, as these are important parameters for the routing phase. The destination table is utilized when a node wants to either locate where the destination node is for its own packet or when it is an intermediate node and wants to route the packet to a specified location. The structures for the neighbor and destination tables are described in Section 5.3.1.2 and Section 5.3.1.3.

Node address	Node latitude	Node longitude	Timestamp	Battery level

FIGURE 5.9
Flooding packet format.

5.3.1.2 Phase II: Network Maintenance

The distinctiveness of proactive routing protocol is that there is the transmission of periodic updates from the nodes. In CDR, these packets are not sent periodically; rather they are sent when there is a change of energy state in the node. The energy state of a node is designated by 3 bits; hence there are eight states for the node's battery level. Once the battery level drops from one state to another, a HELLO packet is transmitted to update the information in the neighboring nodes' tables. This is to minimize the number of transmissions, as the major drawback of proactive systems is that continuous updates drain the battery; updates are only transmitted intermittently and once there is significant change. The node also transmits to its neighbors the number of packets it has received. This will be used later on in the routing phase. During this phase, nodes go to sleep mode to preserve energy and only wake up in periodic intervals for transmitting or receiving packets. The HELLO packet structure is shown in Figure 5.10.

5.3.1.3 Phase III: Routing

The routing phase is essentially where all the decisions are made in regard to disseminating the data and identifying routes and next hop nodes. In this stage, all the information gathered and stored in the two previous phases is utilized to aid with the routing of data packets. Before the routing commences, the data packet header format has to be initialized, as shown in Figure 5.11.

Neighbor address	Neighbor latitude	Neighbor longitude	Type	Timestamp	Battery level	Traffic received

FIGURE 5.10
HELLO packet structure.

Destination address	Priority	Previous distance to destination	Destination latitude	Destination longitude	Timestamp	List of nodes traversed

FIGURE 5.11
Data header format.

The nodes traversed pointer is very important to prevent looping in the network, as this list is used to eliminate entries in the neighbor table from being designated as the next hop.

When a data packet arrives at a node, it is processed to determine what action should be taken. First of all, the node checks the destination address to make sure that it is not the intended recipient. Then the node scans its destination table to make sure that the desired destination does indeed exist and is valid; otherwise, the packet is discarded. If the node turns out to be an intermediate node, or if the node is the source node in transmitting the data packet, the next step would be to determine the next hop. In determining the next hop, first the node clears all expired entries in the neighbor table in order to avoid expired routes. After that, the nodes traversed list is obtained to eliminate any previous nodes to avoid looping in the network. Then the node checks its neighbor table in case one of its neighbors is in fact the destination.

When these steps are completed, the next phase would be to select out of the remaining neighbors the best next hop. The node that is chosen as the best next hop is designated as the priority node. The selection of the priority node is explored in detail in Section 5.4.1. Having three priority nodes will improve the protocol's robustness and decrease the chance of having the rest of the neighbors rebroadcasting, hence decreasing the number of transmitting nodes. Then the packet would be broadcast to all of its neighbors with one or more set as the priority to retransmit.

The next step is how to handle an incoming data packet. Upon receiving a handle to the packet, the node checks the data header to determine if it is the priority neighbor. If it turns out to be the priority, it determines the next priority addresses and rebroadcasts the message to all its next hop neighbors. When this task is completed, it immediately broadcasts an ACK message in order to alert and notify the other nonpriority nodes that are currently in a timeout process. That is because if the node determines from the header that it is not the priority node, it triggers a timeout mechanism where it waits for a specific amount of time for an ACK packet from the priority node. If the node receives the ACK packet within the timeout, it drops the data packet it has received; if, however, the timeout expires and it has not received an ACK packet, it will then proceed to retransmit the data packet. This is intended to ensure reliability, robustness, and fault tolerance in case of a failed transmission or a node failure.

5.3.2 Implementation in OPNET Modeler 15.0

The protocol was chosen to be implemented in OPNET Modeler 15.0™. OPNET Modeler 15.0 can analyze and compare simulated networks in different environments to view the end-to-end behavior of different technology designs and has a development environment to model a wide range of network types and technologies. It allows for total control and various choices in modeling, especially for cross-layer designs or cognitive techniques,

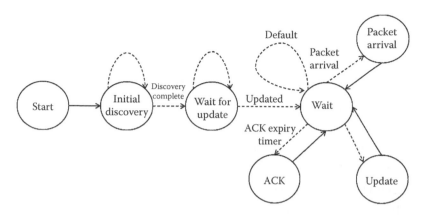

FIGURE 5.12
Process model of CDR.

as it permits the user to design the protocol in every layer. OPNET also is equipped with a detailed and robust discrete event simulator to collect network statistics in order to fully examine and evaluate the performance of a system. This tool allows the measurement of performance metrics on a large-scale network level, to an individual node level, and even to every transmitted packet in the network. The programming language of choice in OPNET is C++, which is important as it permits future migration of the work to a hardware level or to any other program because C++ is one of the most popular programming languages.

The protocol was implemented in the process model, which defines the behavior of a module through finite state machines (FSMs). FSMs utilize states and transitions to determine the actions a module should undertake in response to an event. The C++ code fragments are then attached to each part of the FSM and these code fragments specify the functions in detail to be undertaken when an event occurs. When an event occurs that has an effect on the module, the simulation kernel passes control to the process model through an interrupt. Then the process model responds to the event by transition between states and executing the embedded C++ fragment codes, and then returns the control to the simulation kernel. The process model for CDR is presented in Figure 5.12.

The process model and the FSMs implement the three phases of CDR presented above. The initial discovery FSM is the initial discovery phase, while the network maintenance and routing phases are merged in the remaining FSMs. This model offers a seamless and logical organization of the protocol to enable its operation.

5.3.3 Pseudo-Code for Cognitive Diversity Routing

This section provides a pseudo-code for the above protocol that was implemented in OPNET.

#Start Process

Initialize node

Set battery level

Initialize routing tables

Reset traffic and all counters to zero

Obtains an address based on its ID

Obtains position from GPS (if available)

Set transmission power

#Initial Discovery Process

Node obtains battery level

Flooding packet is formed

Inserts battery and position information into flooding packet

Inserts address and timestamp into flooding packet

Broadcasts flooding packet

Calculate transmission cost

Update battery

Update packet transmission statistic

If flooding packet received

   Process received packet

   If timestamp has expired

      Drop packet

   Else

         Obtain address of the source node

         Compare address to existing addresses in routing table

         Set neighbors in neighbor table and the rest in destination table

         If address does not match addresses in table

            Add new row in table for the extracted address

         Else

            Go to existing row for address

         Extract battery and location information

         Add information to routing table for the relevant address

         From received power find the shadowing variance

         Update table with the channel information

         Calculate the received power from the packet

         Update battery

#Wait Process

Node enters wait mode

If there is a change in energy level

    Initialize update packet and set packet type to Hello

    Insert address into packet header in addition to location information

    Obtain number of packets received from counter and add to packet

    Measure new battery and add to packet

    Insert timestamp at end of packet

    Transmit packet to neighbors

    Calculate transmission cost and update battery

    Update packet statistics

If there is data to transmit

    Initialize data packet and set packet type to Data

    Insert address and location information into header

    Obtain destination information from destination table

    Insert destination address and location into packet

    While iterating through all entries in neighbor table

        If address of entry is in traversed list, ignore entry

        Else

            Obtain energy, traffic and channel info of entry

            Calculate priority factor

            Calculate distance to destination

    Find highest, 2nd and 3rd highest priority factors in entries

    Set the three priorities in packet header

    Calculate distance to destination for determining relay region

    Add node to list of nodes traversed and append list to packet

    Insert timestamp at end of packet

    Transmit packet

    Calculate transmission cost and update battery

    Update packet statistics

If packet is received

    Process packet and obtain type

    If packet type Hello

        Obtain address from packet header

        If timestamp has expired

        Drop packet

Else

    Obtain address of the source node

    Compare address to existing addresses in routing table

    Set neighbors in neighbor table and the rest in destination table

    If address does not match addresses in table

        Add new row in table for the extracted address

    Else

        Go to existing row for address

    Extract battery and location information

    Add information to routing table for the relevant address

    Calculate the received power from the packet

    From received power find the shadowing variance

    Update table with the channel information

    Calculate reception cost

    Update battery

If packet data packet

    Process packet and obtain destination information

    If node is destination

        Obtain info, update statistics and battery then destroy packet

    Node is an intermediate node

    If destination does not exist destination table

        Drop packet

    If address is present in list of traversed nodes

        Drop packet

    Obtain priority addresses from packet header

    If node is first priority

        Go to data transmit function

        Transmit ACK

        Update battery and statistics

    If node is 2nd priority

        Wait timeout T

        If ACK is received

          Drop packet

        Else

          Go to data transmit function

          Transmit ACK

          Update battery and statistics
      If node is 3rd priority
          Wait timeout 2xT
          If ACK is received
            Drop packet
          Else
            Go to data transmit function
            Transmit ACK
            Update battery and statistics
      Else
          Wait timeout 3xT
          If ACK is received
            Drop packet
          Else
            Go to data transmit function
            Transmit ACK
            Update battery and statistics
Else
   Node enters sleep cycle
   Energy consumption is calculated
   Battery updated for every day

5.4 Priority Node Selection

The selection criteria for choosing the next best hop, or the priority node, are based on three parameters or profiles: the energy profile, the channel profile, and the traffic profile. The profiles represent all the considerations that a node needs to take into account before making an intelligent decision on the best next hop.

The energy profile is composed of the energy levels of the receiving nodes and the energy level of the transmitting node itself. The node iterates through its neighbor table to determine the best receiving node. This is done by subtracting the estimate reception cost P_{R0} from the neighbor node battery, P_{NBR}, obtained from the neighbor table. This value will represent the estimated battery of the neighbor node should it receive the packet, which is denoted as P_{NBR-E}:

$$P_{NBR-E} = P_{NBR} - P_{R0}$$

Then for each of these neighbors, the node estimates the transmission cost P_{T-E} to this specific neighbor by subtracting the estimated transmission cost $P_T(d)$ from the node's current battery P_N.

$$P_{T-E} = P_N - P_T(d) = P_N - P_{T0} - A \times d^\alpha$$

where A and P_{T0} are constants (see Section 5.2.1) (Kim B. and Kim I. 2006; Wang and Yang 2007; Molish et al. 2009).

Combining both estimated batteries for transmission and reception yields the equation for the energy profile, which is equal to:

$$E_P = P_{T-E} + P_{NBR-E} = P_N + P_{NBR} - P_{R0} - P_{T0} - A \times d^\alpha$$

The channel profile provides the node with information about the channel in order for the node to consider the condition of the link. The shadowing effect is vital to consider when evaluating a link because it identifies any obstacles that might hamper the transmission. Based on the propagation model, by obtaining the received power from a packet transmitted on the link, the shadowing effect, and thus the channel profile C_p between a transmitter i and a receiver j, is represented by the variance of the lognormal shadowing variable X_σ:

$$C_p = \sigma_{i,j}^2$$

Finally the last profile to be considered is the traffic profile. The traffic profile provides statistics about the number of packets that a node is receiving. This parameter is used to avoid congestion and potential bottlenecks, as sometimes several nodes end up transmitting to the same node, causing collisions and energy stress at that node. Also, in random deployments, one node may have the choice while a forced path might be the only choice for other nodes; hence this profile will help avoid transmitting to these nodes. Trb_{NBR} is the number of packets a neighbor has received, which is obtained from the neighbor table. It is divided by the total number of packets estimated in the entire network, which is equal to

$$btr \times t \times K$$

where btr is the transmission rate, t is the time elapsed, and K is the number of nodes in the network.

The traffic profile Tr_p is:

$$Tr_p = \frac{Trb_{NBR}}{btr \times t \times K}$$

Therefore, combining all three profiles yields the following equation to determine the priority node:

$$Priority_N = k_1 \times E_P - k_2 \times C_p - k_3 \times Tr_p$$

$$Priority_N = k_1 \times (P_N + P_{NBR} - P_{R0} - P_{T0} - A \times d^\alpha) - k_2 \times \sigma^2 - k_3 \times \left(\frac{Trb_{NBR}}{btr \times t \times K} \right)$$

where k_1, k_2, & k_3 are weights assigned to each profile.

Combining those three profiles enables the node to possess the latest updated information about its energy level, its neighbor's energy level, the condition of the link, and the traffic distribution in the network. After obtaining the equation for priority selection, the next step would be to determine the value of the three weights k_1, k_2, & k_3 in order to maximize network lifetime and optimize the selection process to choose the best priority node. Therefore, the selection needs to be formulated as an optimization problem according to the designer's requirements. For example, the designer can put more emphasis on the energy, or on the channel profile, and so forth.

5.5 Performance Evaluation

In this section, the cognitive diversity routing (CDR) protocol is tested using the simulation tool of OPNET Modeler 15.0. CDR is simulated versus the geographic random forwarding protocol (GeRaF) in order to measure the efficiency and advantages of CDR. GeRaF is a widely used location-based routing protocol that utilizes a node's GPS capabilities to select a route based on distance from the destination. Both protocols were tested under identical conditions in different scenarios. Different scenarios were chosen in order to test how both protocols performed under different conditions and for different applications, as that will yield an overview of the strengths and weaknesses of the protocols.

The deployment areas in the scenarios were chosen to be 1.5×1.5 km in size and the transmission range of each node is chosen to be approximately 500 meters. For each scenario, the transmission rate was varied between 1 and 24 transmissions per day; that is, the sensor node transmits every hour. This is to simulate the networks under varying traffic intensities. Traffic was

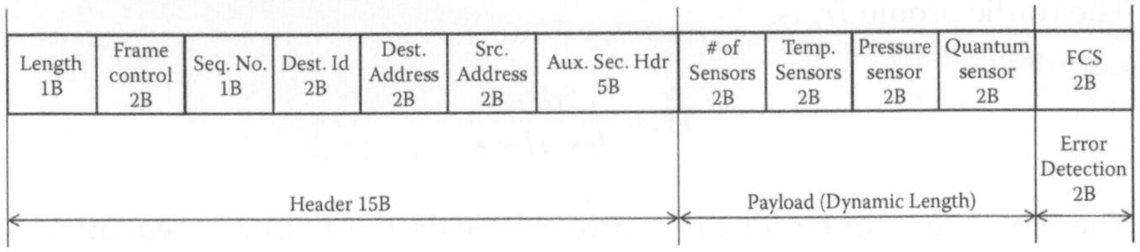

Length 1B	Frame control 2B	Seq. No. 2B	Dest. Id 1B	Dest. Address 2B	Src. Address 2B	Aux. Sec. Hdr 5B	# of Sensors 2B	Temp. Sensors 2B	Pressure sensor 2B	Quantum sensor 2B	FCS 2B

Header 15B   ←→   Payload (Dynamic Length)   ←→ Error Detection 2B

FIGURE 5.13
Packet format.

generated by each of the transmitting nodes at the defined rate per day, and all the data were directed toward the sink node.

The modeled traffic in the network was chosen to represent the traffic in an environment monitoring application (Ibnkahla 2010), so the packet format was given as shown in Figure 5.13. The size of each packet was chosen to be a random value between 136 and 200 bits, which represents 15 to 25 bytes according to the packet format given above. The range signifies the number of sensor readings that are being transmitted to the sink node. The path loss exponent, unless specified otherwise, is taken to be 3, as that represents the typical and practical environment for this type of application (see Table 5.1).

In the simulations below, the three weights of the objective function were optimized by giving emphasis to the energy profile (energy profile optimization). An optimized method yielded the values: $k_1 = 8.57$, $k_2 = 5.71$, and $k_3 = 5.71$.

The two main metrics for the simulations are the first node death (FND) and the network lifetime. FND is vital in monitoring applications where each node is important, and the goal is to extend the lifetime of each node as it relays information that no other node can compensate. FND also helps illustrate the cooperative routing and the diversity that CDR protocol carries out, and therefore is an important metric in characterizing a protocol. The network lifetime metric is the main objective, as the purpose of developing the CDR protocol is to extend the network lifetime. The environment network lifetime definition explained earlier is taken as the network lifetime definition, and the threshold is taken to be 40% or when the network is partitioned such as a set of nodes will be unable to reach the sink node. The simulations

TABLE 5.1

Physical Parameters for Scenario Testing

Parameter	Value
Area	1.5×1.5 km
Transmission Range of Node	500 meters
Packet Size	136–200 bits
Modulation	DPSK
Noise Figure	1.0
Transmitter Bandwidth	22.0 kHz
Path Loss Exponent α	$\alpha = 3$

were carried out on three different deployment scenarios: the grid deployment, the forced path deployment, and the random deployment.

5.5.1 Grid Deployment

The first scenario has been chosen to be 1.5×1.5 km square area, and 21 nodes have been placed in a grid formation. Twenty of the 21 nodes are transmitting nodes, while the other node is the sink node to which all traffic is directed. The grid deployment is a great way to demonstrate the strengths of a protocol, as there are many different paths to the sink node; thus it allows a protocol to fully display its array of flexibility in choosing the optimal path from the source node to the destination. The nodes are arranged in the grid as displayed in Figure 5.14.

The purpose of the first simulation is to calculate the first node death (FND) and network lifetime for both protocols under the specified conditions.

As Figure 5.15 shows, CDR's FND is significantly higher than GeRaF's. This is due to the cognitive aspect of the routing in CDR. In GeRaF, all routing decisions were made based on static parameters; hence the protocol was not able to adapt to changes in the network and environment. Nodes that were at the beginning of the simulation regarded as the best forwarding nodes have been overloaded with traffic, thus causing them to die quickly. On the other hand, CDR is adapting to changes in the network and evenly distributing the traffic load, resulting in an even distribution of energy losses throughout the network. The network lifetime measured is displayed in Figure 5.16.

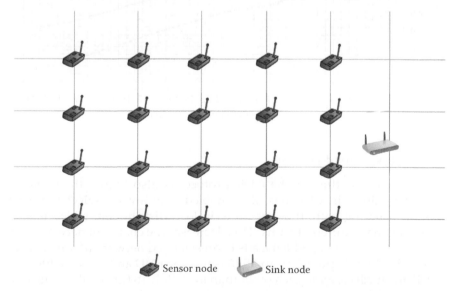

Sensor node   Sink node

FIGURE 5.14
Grid deployment.

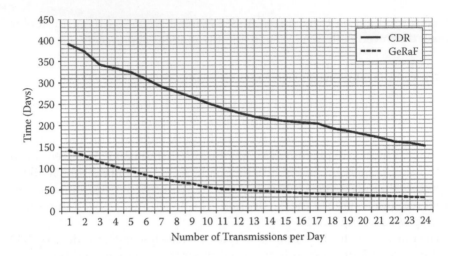

FIGURE 5.15
FND for grid deployment.

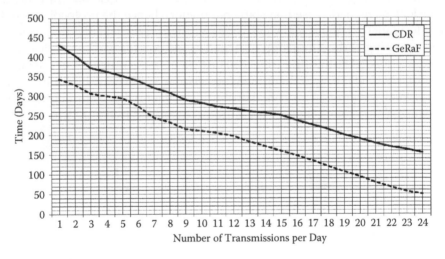

FIGURE 5.16
Network lifetime for grid deployment.

The network lifetime of the CDR protocol is also higher than that of GeRaF. Cognition in CDR has also enabled it to always select the least energy-consuming path, thus resulting in load distribution, which results in selecting many different paths. This decreases the chance of network portioning and segregation, which leads to an extended network lifetime. What is noticeable from Figures 5.15 and 5.16 is that the FND and network lifetime in CDR are much closer to each other than in GeRaF. To further illustrate that point, the FND and network lifetime are plotted together on the same graphs for each protocol, as displayed in Figure 5.17 and Figure 5.18.

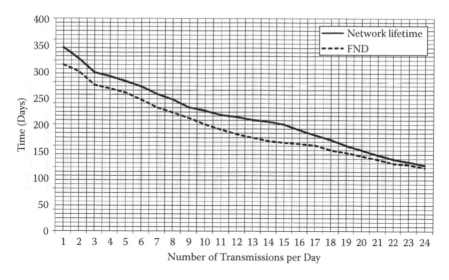

FIGURE 5.17
CDR network lifetime vs. FND.

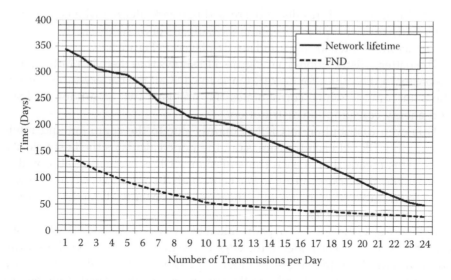

FIGURE 5.18
GeRaF network lifetime vs. FND.

Figure 5.17 and Figure 5.18 prove the previous assertion that the gap between the network lifetime and FND in CDR is much less than in GeRaF. This also validates the previous statement about load balancing in CDR, where it appears that the first node death is only 8–22% away from the network life-time, while in GeRaF, that range can go up to 188%. This polarity leads to congestion in traffic, as it is always directed through a narrow path and that leads

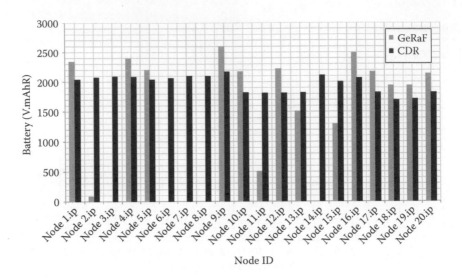

FIGURE 5.19
Battery distribution after 100 days.

to partitioning, which results in decreased network lifetime when compared to CDR. To further explore this point, the batteries of each node in the network were measured after 100 days and are displayed in Figure 5.19.

The battery distribution from Figure 5.19 shows that there is a rather large disparity in the values for GeRaF, while in CDR all the battery values are in the same range—which corroborates the previous two assertions on load and energy-consumption distribution.

This even distribution in CDR is mainly due to the cognition and intelligence in CDR. At each transmission, the node calculates the best priority node. This means that at each transmission, CDR is evaluating all possible options and then transmitting to least cost path. In order to verify this assertion, the number of paths taken is measured in order to show the variation between both protocols:

In Figure 5.20, the number of paths taken from each of the nodes to the destination is recorded. The nodes at the edge of the grid and placed farthest away from the sink have the largest number of possible paths, as there are many possibilities from the end nodes to the sink node. As mentioned at the beginning of this section, the main advantages of grid deployments are the number and variety of paths and options it offers a source node to reach the destination, and that plays to the advantage of CDR, which attempts to find the least-cost path. Therefore, the larger the number of paths available, the more options it will have, which results in choosing better paths that decrease the overuse of those nodes that were traversed by most source nodes. Since the cognition aspect involves an energy profile, a channel profile, and a traffic profile, the path can easily navigate by avoiding congestion, bad channel conditions, and

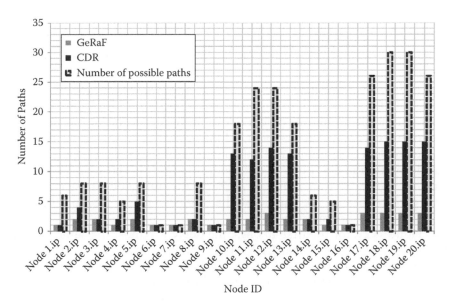

FIGURE 5.20
Numbers of paths to destination.

nodes that have low batteries. However, one thing that might raise a bit of concern from Figure 5.20 is that the larger number of paths and its impacts on the number of hops and delay it might cause. Therefore, the number of hops from source to destination for each node was measured and is presented in the graph shown in Figure 5.21.

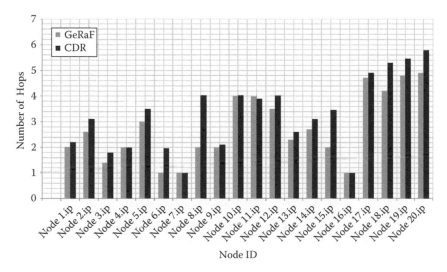

FIGURE 5.21
Number of hops to sink node.

Figure 5.21 shows that even though Figure 5.20 displayed that the number of paths is significantly larger, the number of hops is not that much greater. This is important because it dispels one extreme case of the CDR protocol whereby it takes very long paths and lots of hops to reach the destination, thus negating its objective of minimizing energy consumption in the network. In CDR, the number of hops is a trace larger than in GeRaF. Sometimes it is better to avoid intermediate nodes that have significant congestion or where channel conditions are poor; taking an extra hop is thereby justified.

One problem that might arise from longer paths and extra hops is end-to-end delay. Two nodes, node 17 and node 19, were chosen to measure their end-to-end delay in transmitting a packet to the sink node. Those two nodes were chosen because they are located at the edge of the grid and have the largest path and hop numbers.

In Figure 5.22, for the first 40 days, the delay is almost the same for both protocols; however, after that, CDR would exhibit a delay of an additional 0.1 to 0.15 seconds. For the first 40 days, all the surrounding nodes of node 17 were decreasing in energy at almost the same level, so the paths that CDR and GeRaF took were very similar; however, after the network had been alive for a while and the traffic and energy values distorted through operation, CDR takes a longer path to avoid congestion zones and nodes in the center that have heavy loads—so the delay increases for CDR. On the 89th day, there is an additional 0.05 second delay. That is due to diversity routing, where the priority node selected was unable to transmit; after waiting for a 0.05 second timeout, the next priority node transmits, causing an overall additional delay.

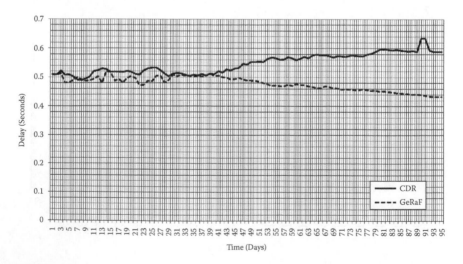

FIGURE 5.22
Delay from node 17 to sink.

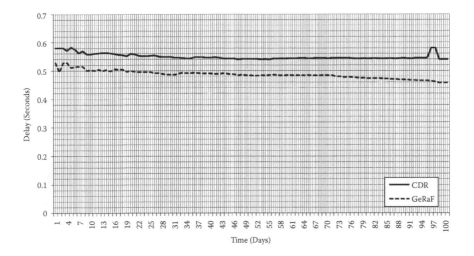

FIGURE 5.23
Delay from node 19 to sink.

In Figure 5.23, the delay of CDR is consistently around 0.1 second larger than the delay in GeRaF. That is because after the initialization phase, the shortest path selected by GeRaF passes through all the central nodes where congestion is occurring; the path that CDR chose to avoid those nodes took extra hops, which resulted in the additional delay.

Therefore, even though the CDR takes more paths with a larger hop count than the GeRaF protocol, it still manages to avoid the two main disadvantages of that process. The first is that the larger number of hops means more nodes are transmitting. More energy consumed is disproven because, as it has been established and demonstrated, distributing the load results in the entire consumption cost being distributed over the entire network, thus resulting in a longer network lifetime. The second disadvantage is the increase in end-to-end delay, where the delay incurred through extra hops and additional processing power accounts for no more than 26% of the optimal delay, which is a completely acceptable range for the applications in which the main objective is not rapidness or end-to-end delay.

As Figure 5.24 demonstrates, the throughput was almost the same until around 54 days. That signifies the FND for the GeRaF protocol. From day 54 until the death of the network on day 201, the throughput decreases with the death of nodes in the network. Note that the network lifetime for GeRaF at the transmission rate of 12 transmissions per day is 187; in this graph, the throughput is stable for a further period of time. This is due to the definition of the network lifetime where segregation occurs or the death of 40% of the nodes. However, in this case, even when the segregation occurs, there are still some nodes that are able to transmit to the sink node. In CDR, the throughput is consistent for 222 days until the first node death occurs. However, unlike

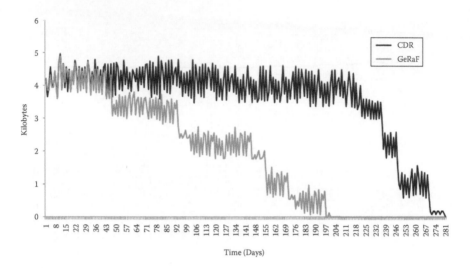

Time (Days)

FIGURE 5.24
Throughput.

GeRaF, in CDR only the dead nodes are unable to transmit, as the diversity routing mechanism doesn't allow packets to be dropped; rather, they are still transmitted to the destination after the timeout mechanism is activated.

5.5.2 Deployment with Forced Path

The previous scenario was the grid deployment where the node had a full array of path choices to make. This resulted in the CDR protocol performing much better than GeRaF because it had many different alternate paths to take in order to maximize the network lifetime. However, while the grid scenario can be classified as one extreme case scenario due to the node having many options, it is also important to test the protocol at the other extreme. The other extreme would be to completely restrict the number of choices or paths the node can make. The following scenario depicts this case, where nodes are clustered into different groups and are only connected to each other via a two-node "bridge," restricting the next hop to two nodes only. The scenario is presented in Figure 5.25.

The same simulation is carried out on this scenario in order to measure the FND (Figure 5.26) and network lifetime (Figure 5.27) for this radical case.

The results shown in Figure 5.27 are entirely predictable: The network lifetime is the same for both protocols. However, results in Figure 5.26 show that the FND for CDR is greater than that for GeRaF. In this scenario, the following process is occurring. Nodes in the first cluster have to transmit through nodes 11 and 18, while nodes in the second cluster have to transmit through nodes 5 and 9. There are only two paths for each cluster, which is why the network lifetime is the same for both protocols, as the network is considered dead once nodes

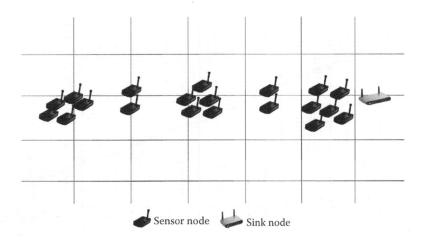

Sensor node   Sink node

FIGURE 5.25
Forced path deployment.

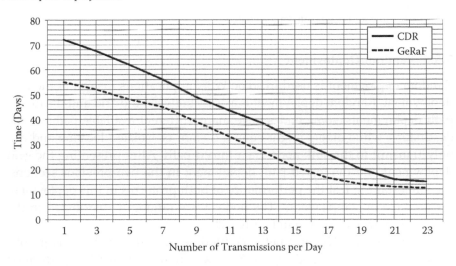

FIGURE 5.26
FND for forced path deployment.

11 and 18 or nodes 5 and 9 die to partition the network. However, the mode of operation of both protocols is visibly evident from the FND graph. GeRaF transmits to the node in the bridge that is closest to it until it dies and then moves to the other node in the bridge, while in CDR, the load is distributed across both nodes so they both die at roughly the same time. The importance of this scenario is that it displays the behavior of both protocols in bottleneck cases.

Not all WSN applications are deployed in grid fashion; some are deployed in a random distribution. In random distributions, small clusters and forced paths might occur, hence the need to consider this case.

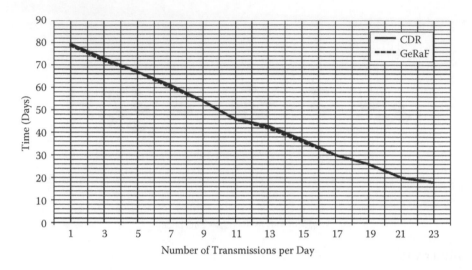

FIGURE 5.27
Network lifetime for forced path deployment.

5.5.3 Random Deployment

Random deployment is used in some WSN applications instead of the grid deployment. Random deployment is a cross between the grid and the forced path deployments; some nodes have many different available paths to the destination while others might be in a distant or isolated location and can only transmit to the sink node via a forced path. After presenting the grid deployment and the forced path deployment in the two previous sections, it is important to test the performance of both protocols in this deployment for the sake of completeness. The nodes were randomly distributed in the 1.5 × 1.5 km grid, as shown in Figure 5.28.

The results for FND and network lifetime are presented in the graphs shown in Figure 5.29 and Figure 5.30, which show that CDR performed much better than GeRaF, which was to be expected after the results of the grid deployment and the forced path deployment. The network lifetime in the random deployment is also around 103 days less than that of the grid deployment. In the grid deployment, all nodes have several paths to choose from, while in random deployment some nodes have only forced paths, hence decreasing network lifetime. (The FND for GeRaF in Figure 5.29 is also much less than the one for CDR due to the same reason as Figure 5.26 in the forced path deployment.) Therefore, in random deployment networks, CDR also performs much more admirably than GeRaF because it takes advantage of the nodes with several path options to choose the least-cost path, thus extending network lifetime.

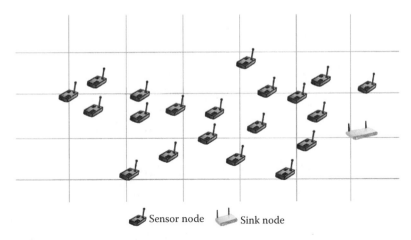

Sensor node   Sink node

FIGURE 5.28
Random deployment.

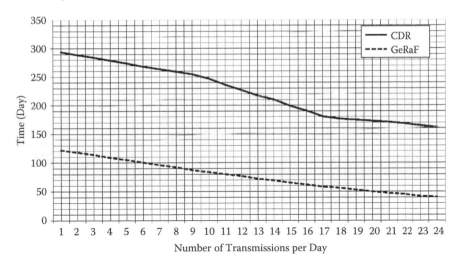

FIGURE 5.29
FND for random deployment.

5.5.4 Node Density and Scalability

All of the previous scenarios have featured different deployments but the same number of nodes (20 nodes). Increasing the node density or the number of nodes is a good way of characterizing a protocol, as it shows its scalability. The first aspect resulting from increasing the number of nodes would be that the traffic in the network would increase, resulting in an increase in the chance of congestion. The other aspect is that the number of possible paths in the network would increase because each additional node is a new candidate

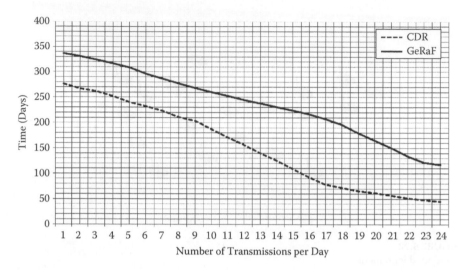

FIGURE 5.30
Network lifetime for random deployment.

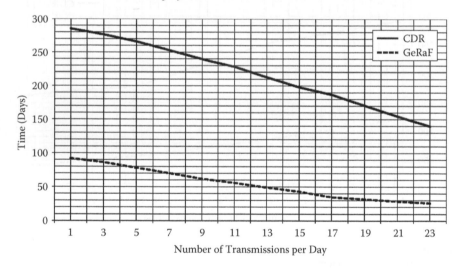

FIGURE 5.31
FND for 30 nodes in grid deployment.

for a path. Therefore, the number of nodes in the network was increased to 30 and 40 nodes, and the network lifetime and FND were measured.

5.5.4.1 Grid Deployment

In the first scenario, the nodes were arranged in a grid deployment. The grid deployment allows for a high variation of path options from the source node to the sink.

It can be concluded from Figure 5.31, Figure 5.32, Figure 5.33, and Figure 5.34 that CDR performs better than GeRaF: The network density increases, proving that it is a robust option for larger networks. Although the network lifetime of GeRaF degrades quickly as the network density increases in CDR, it only decreases slightly. This is because of the different effects that the two aspects of increasing node density have on the two respective protocols. GeRaF suffers because there are more nodes, more traffic, and more congestion, which leads to a serious degradation in its network lifetime.

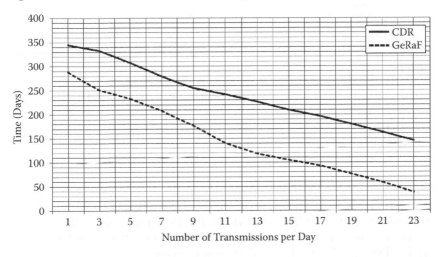

FIGURE 5.32
Network lifetime for 30 nodes in grid deployment.

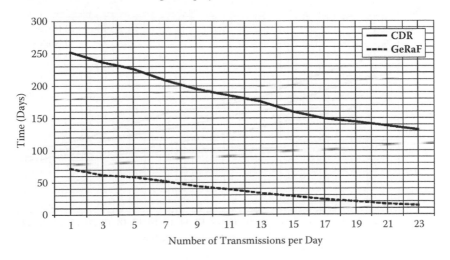

FIGURE 5.33
FND for 40 nodes in grid deployment.

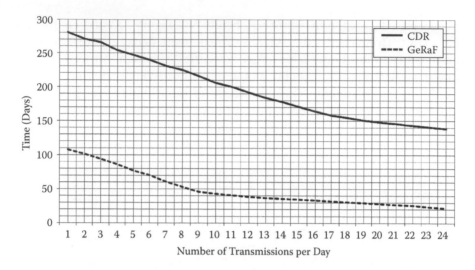

FIGURE 5.34
Network lifetime for 40 nodes in grid deployment.

On the other hand, CDR takes advantage of the second aspect, which is an increase in the choice of routes, to offset the disadvantage of having increased traffic and congestion in the network. To put that into more context, the performances of the three different node densities for CDR and GeRAF are displayed in Figure 5.35 and Figure 5.36, respectively.

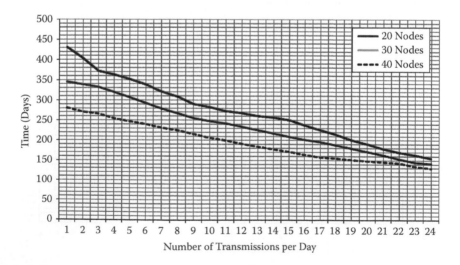

FIGURE 5.35
Network lifetime scalability comparisons for CDR.

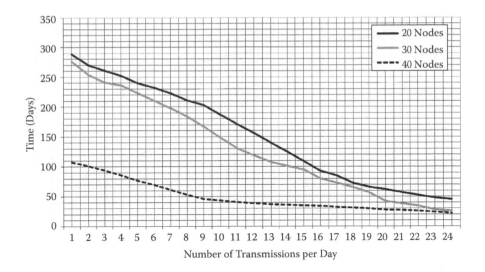

FIGURE 5.36
Network lifetime scalability comparisons for GeRaF.

5.5.4.2 *Random Deployment*

The scalability of both protocols was then tested on a random deployment. Random deployment restricts the number of paths and some nodes have forced paths to reach the sink, so segregation and node isolation have a higher probability of occurring (see Figure 5.37, Figure 5.38, Figure 5.39, and Figure 5.40).

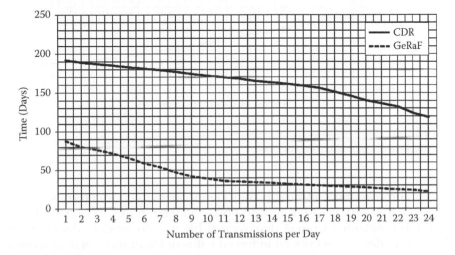

FIGURE 5.37
FND for 30 nodes in random deployment.

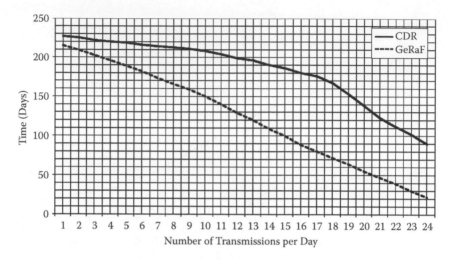

FIGURE 5.38

Network lifetime for 30 nodes in random deployment.

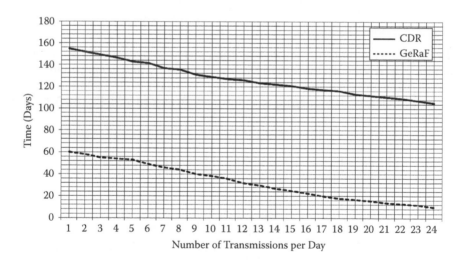

FIGURE 5.39

FND for 40 nodes in random deployment.

CDR performs more admirably than GeRaF for both node densities, 30 and 40 nodes. The network lifetime and FND obtained in this section are lower than those obtained in grid deployment because in random deployment, there are nodes that get placed in forced paths or the number of neighbors is less than the grid deployment, restricting the number of path options available for routing.

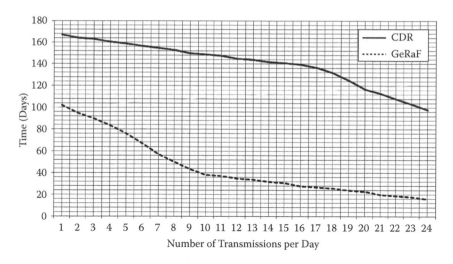

FIGURE 5.40
Network lifetime for 40 nodes in random deployment.

5.5.5 Optimization

In the previous simulations, the three weights were assigned the optimized values $k_1 = 8.57$, $k_2 = 5.71$, and $k_3 = 5.71$ that gave emphasis on the energy profile.

The simulation below was conducted to compare the above weights to a nonoptimal assignment of weights by setting the value of k_1, k_2, and k_3 to 1. The chosen scenario was the random deployment distribution.

Figure 5.41 indicates that in low number of transmissions per day, there is not much difference; however, as the traffic increases in the network, the optimized case outperforms the nonoptimized case. This is because when there is not a greater emphasis on the energy profile; nodes might die after being selected as the priority instead of avoiding them. When there is no emphasis on the energy and traffic profile that might result in the elimination of a node, that is an enforced path for other nodes. For example, a node might have two options: one is a viable option and the other is an enforced path for another node; when traffic and energy profiles are not considered, the node loses a sense of perspective of the surrounding environment. Also, if the entire emphasis is placed on the energy profile, for example, the node will be unable to avoid bad channel conditions; hence the priority node will not be able to receive the packet due to the channel causing the packet to be below the signal to noise ratio (SNR) and received power threshold. That will lead to either a retransmission or a dropped packet. This scenario might also lead to traffic congestion at enforced paths. By assigning the optimal weights, improved performance in the network is ensured.

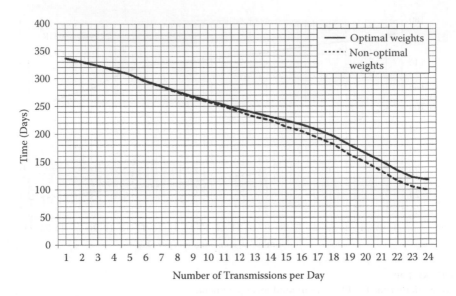

FIGURE 5.41
Optimized vs. non-optimized weights results.

5.5.6 Giving Emphasis to the Channel Profile

This section studies the scenario where the priority node selection problem is solved when the channel profile is given priority (channel optimization). The optimal priority node selection objective function weights were obtained as: $k_1 = 8.89$, $k_2 = 6.67$, and $k_3 = 4.44$.

This scenario occurs when the channel profile needs to be given priority over the energy profile in order to show the impact of the optimization. The shadowing and path loss variables were changed and increased in order to portray links with severe fading due to shadowing, thus causing importance to be placed on the channel profile. Measuring network lifetime and first node death will not be the primary concern, since when the channel conditions are extreme, packets will be dropped and will not reach their destination. In order to measure reliability, the number of dropped bytes was measured. It is important to note that in CDR, every node drops the received packet if it is not a priority node and it receives an ACK; therefore, packets that were dropped due to the channel conditions were marked and measured. Figure 5.42 and Figure 5.43 present the results that were taken from a grid deployment:

CDR is more reliable as the number of bytes dropped is much smaller than that of the other protocol. This is due to CDR choosing paths with the best channel conditions, avoiding paths with high shadowing, thus decreasing the number of packets dropped and lost. Decreasing the number of packets dropped will also decrease transmissions; if a transmission is unsuccessful,

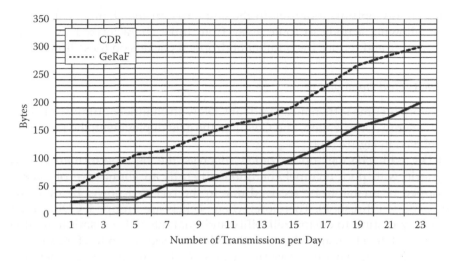

FIGURE 5.42
Number of bytes dropped.

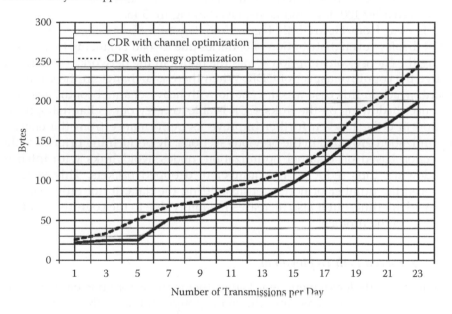

FIGURE 5.43
Number of bytes dropped due to optimization.

then a second- or third-priority node will have to retransmit the information, which will cause more energy consumption in the network, thus reducing network lifetime. Figure 5.43 displays CDR performance with channel optimization as opposed to the energy optimization. Since the channel conditions are severe, the CDR with channel optimization outperforms the CDR with energy optimization due to the emphasis on the channel conditions, thus allowing a node to route away from links characterized by intense shadowing and reduce the number of dropped packets on these links.

5.5.6.1 Grid Deployment

Network lifetime is then measured for CDR with channel optimization, CDR with energy optimization, and GeRaF in the grid deployment; the results are displayed in Figure 5.44.

CDR with channel optimization displays the best performance. In a scenario with poor channel conditions, there is an abundance of packets dropped or unable to be received due to the received power being below the expected threshold. Therefore, when packets are dropped, CDR utilizes diversity routing to ensure reliability and guarantee of transmission; more nodes will then transmit (the second- and third-priority nodes, for example). Hence when more nodes transmit, the network lifetime decreases; therefore, the protocol that increases reliability also increases network lifetime because it decreases the number of retransmitting nodes. The battery for each node is measured after 100 days and is displayed in Figure 5.45.

It is clear from Figure 5.45, Figure 5.46, Figure 5.47, and Figure 5.48 that CDR displays even distribution of batteries due to load balancing by distributing the energy consumption across the nodes in the network. The figures prove the cognitive aspect of CDR. When there is a change in the environment, which in this case was a scenario with heavy channel conditions and intense shadowing on the links, CDR is able to carry out a channel optimization in order to even further improve its performance from the energy optimization provided in the previous sections. By performing priority node selection based on channel optimization, nodes avoid links with intense shadowing, thus reducing the number of dropped packets. This in turn reduced the number of nodes retransmitting the data. It not only extends network lifetime, but it also increases reliability due to the cognitive characteristics of the protocol.

5.5.6.2 Forced Path Deployment

The forced path deployment depicts the other extreme for node path options, where nodes are clustered into different groups and are only connected to each other via a two-node "bridge," restricting the path choice for the next hop to two nodes only. The network lifetime was measured for the three protocols:

Figure 5.49 exhibits a different behavior than that shown in Figure 5.27. In Figure 5.27, the network lifetime for both protocols was very similar; however, here there is a difference between each protocol. In a scenario with heavy channel conditions, a larger number of packets are dropped, which leads to more retransmissions, which in turn reduces the network lifetime. In the scenario of Figure 5.27, the network lifetime depended on the number

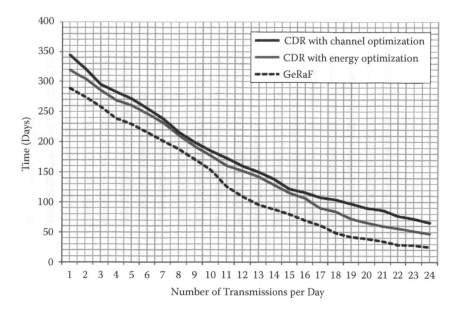

FIGURE 5.44
Network lifetime for channel optimization.

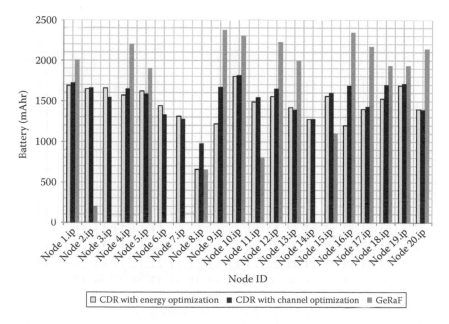

FIGURE 5.45
Protocol comparison for battery distribution after 100 days.

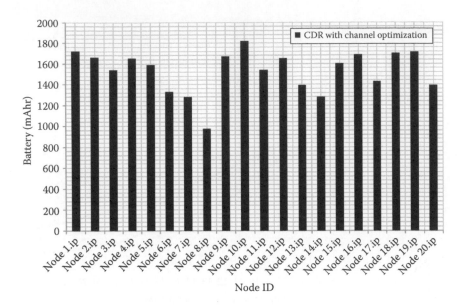

FIGURE 5.46
Battery distribution after 100 days for CDR with channel optimization.

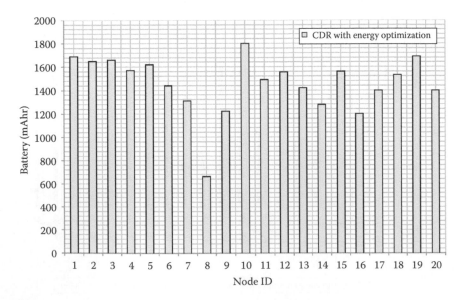

FIGURE 5.47
Battery distribution after 100 days for CDR with energy optimization.

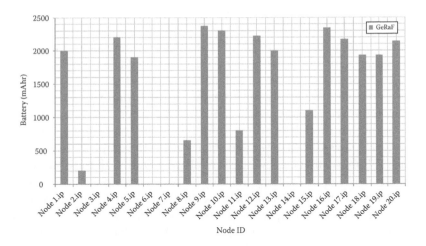

FIGURE 5.48
Battery distribution after 100 days for GeRaF.

of paths the protocol took; however, in the case of Figure 5.49, the network lifetime is dependent on the reliability of the protocol, which in turn affects the network lifetime. This explains why CDR with channel optimization is the best performer and GeRaF has the lowest network lifetime.

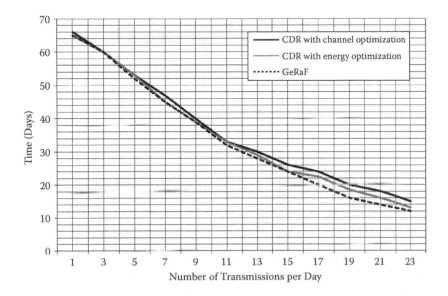

FIGURE 5.49
Network lifetime for forced path deployment.

5.5.6.3 Random Deployment

Random deployment combines both the grid and the forced path deployments, where some nodes have many different available paths to the destination while others might be in a distant or isolated location and can only transmit to the sink node via a forced path. Network lifetime is expected to be less than in the grid deployment because several nodes will die more quickly due to the higher probability of network partition. The network lifetime for the three protocols is measured and displayed in the graph in Figure 5.50.

Figure 5.50 shows that CDR with channel optimization slightly outperforms CDR with energy optimization, and they both outperform GeRaF. That is due to the increase in the number of packets dropped, leading to more retransmissions, which increases energy consumption. Network lifetime is dependent on the reliability of the protocol—which is why CDR with channel optimization is the best performer and GeRaF has the lowest network lifetime.

The previous three scenarios display the cognitive aspect of CDR. When there is a change in the environment, such as a scenario with heavy channel conditions and intense shadowing on the links, CDR is able to carry out a channel optimization in order to even further improve its performance from the energy optimization described in the previous sections. By performing priority node selection based on channel optimization, nodes avoid links with intense shadowing, thus reducing the number of dropped packets. This in turn reduces the number of nodes retransmitting the data, not only

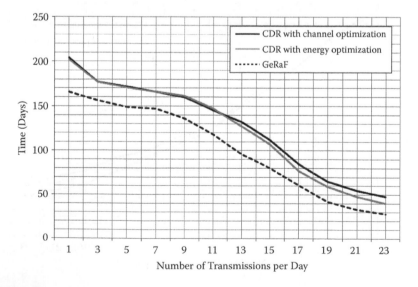

FIGURE 5.50
Network lifetime for random deployment.

extending network lifetime but also increasing reliability due to the cognitive characteristics of the protocol. This displays the node's ability to adapt intelligently to changes in the network, characterizing the CDR protocol as a cognitive protocol.

5.6 Conclusion

This chapter presented a cognitive diversity routing protocol for wireless sensor networks with the main objective to extend network lifetime. CDR is a cognitive protocol since it adapts intelligently to network conditions through perception, reasoning, and gathering knowledge. The protocol ensures that the nodes in the network carry out observations regarding the state of the network and share information in order to learn and reason before carrying out optimization decisions. By utilizing the energy, channel, and traffic profiles, the node is aware of all aspects that can affect the energy efficiency of the network, and thus is able to adapt cognitively to dynamic conditions in order to achieve end-to-end network goals. Diversity routing techniques are utilized to increase reliability and decrease the number of unnecessary retransmissions. Simulations and performance evaluations have proven that CDR outperforms another popular protocol by ensuring load balancing to distribute energy consumption in the network. The load balancing is due to the different and larger number of paths the protocol enables routes to take. Nodes transmit to the best next hop at each step, ensuring distributed energy consumption across the network, which reduces network segregation, node isolation, congestion, and bottlenecks. This incurs an additional end-to-end delay; however, the delay is not very significant and remains within tolerable ranges. Through cognition, nodes are able to adapt to changes in the network, whether the changes occur in channel conditions or in traffic congestions spots. This enables the protocol to be robust and scalable; it is able to find an advantage when the network density is higher because that increases the possible number of paths available. CDR also presents improved throughput and increased reliability by employing the notion of diversity routing, presenting more degrees of freedom than GeRaF. CDR performed well in various types of deployments such as grid or random deployments. In the grid deployment, CDR took advantage of the deployment's characteristic by exploiting the large number of paths available in the network; in random deployment, the protocol protected nodes in forced paths or in danger of isolation by routing traffic away from their "bridge nodes" (nodes in the forced path), which resulted in extended network lifetime.

References

Akan, O.B., Karli, O., and Ergul, O. (2009, July–August). Cognitive radio sensor networks. *IEEE Network* 23(4), 34–40.

Al-Karaki, J.N., and Kamal, A.E. (2004). Routing techniques in wireless sensor networks: A survey. *IEEE Wireless Communications*, 6–28.

Al-Turjman, F.M., Hassanein, H.S., and Ibnkahla, M.A. (2009). Connectivity optimization with realistic lifetime constraints for node placement in environmental monitoring. *IEEE 34th Conference on Local Computer Networks, 2009. LCN 2009.* Zurich.

Boonma, P., and Suzuki, J. (2007). Evolutionary constraint-based multiobjective adaptation for self-organizing wireless sensor networks. *2nd Bio-Inspired Models of Network, Information and Computing Systems, 2007. Bionetics 2007,* Budapest, pp. 111–119.

El Mougy, A., Bdira, E., and Ibnkahla, M. (2010). Throughput optimization of a power-aware MAC for WLANs in correlated shadowing environments. *25th Biennial Symposium on Communications (QBSC), 2010,* Kingston, Ontario, pp. 10–13.

El Mougy, A., El-Jabi, Z.H., Ibnkahla, M., and Bdira, E. (2010). Cognitive approaches to routing in wireless sensor networks. *IEEE Global Telecommunications Conference, 2010. GLOBECOM 2010,* Miami, FL.

El-Jabi, Z. (2010). *Cognitive Diversity Routing in Wireless Sensor Networks,* MSc. thesis, Queen's University, Canada.

Felemban, E., Lee, C.-G., and Ekici, E. (2006, June). MMSPEED: Multipath multi-speed protocol for QoS guarantee of reliability and timeline in wireless sensor networks. *IEEE Transactions on Mobile Computing* 5(6), 738–754.

Fortuna, C., and Mohorcic, M. (2009, June). Trends in the development of communication networks: Cognitive networks. *Computer Networks* 53(9), 1354–1376.

Frey, H., Ruhrup, S., and Stojmenovic, I. (2009). Routing in wireless sensor networks. In *Guide to Wireless Sensor Networks,* edited by S.C. Misra, I. Woungang, and S. Misra, 81–111. Paderborn, Germany: Springer London.

Griva, I., Nash, S.G., and Sofer, A. (2009). *Linear and Nonlinear Optimization,* 2nd ed. Fairfax, Va.: Society for Industrial and Applied Mathematics.

Ibnkahla, M. (2010). *Versatile Wireless Sensor Network for Environment Monitoring Applications with High Modularity.* Technical Report, Queen's Wireless Communications and Signal Processing Laboratory, Kingston, Ont.

Kim, B., and Kim, I. (2006, January). Energy aware routing protocols in wireless sensor networks. *International Journal of Computer Science and Network Security* 6(1).

Kuruvila, J., Nayak, A., and Stojmenovic, I. (2006, July). Progress and location based localized power aware routing for ad hoc and sensor wireless networks. *International Journal of Distributed Sensor Networks* 2(2), 147–159.

Lenders, V., and Baumann, R. (2008). Link-diversity routing: A robust routing paradigm for mobile ad hoc networks. *IEEE Wireless Communications and Networking Conference, 2008. WCNC 2008,* Las Vegas, pp. 2585–2590.

Mansouri, V., Ghiassi-Farrokhfal, Y., Nia-Avval, M., and Khalaj, B. (2005). Using a diversity scheme to reduce energy consumption in wireless sensor networks. *2nd International Conference on Broadband Networks, 2005. BroadNets 2005,* Boston, pp. 940–943.

Mizanian, K., Yousefi, H., and Jahangir, A.H. (2009). Modeling and evaluating reliable real-time degree in multi-hop wireless sensor networks. *Sarnoff Symposium, 2009. SARNOFF '09. IEEE*, Princeton, N.J., pp. 1–6.

Molisch, A.F., Greenstein, L.J., and Shafi, M. (2009, May). Propagation issues for cognitive radio. *Proceedings of the IEEE* 97(5), 787–804.

Niezen, G., Hancke, G.P., Rudas, I.J., and Horvath, L. (2007). Comparing wireless sensor network routing protocols. *AFRICON 2007*, Windhoek, pp. 1–7.

Oteafy, S., AboElFotoh, H.M., and Hassanein, H.S. (2009). Dynamic election-based sensing and routing in wireless sensor networks. *IEEE Global Telecommunications Conference, 2009. GLOBECOM 2009*, Honolulu, Hawaii, pp. 1–6.

Ozgovde, A., and Ersoy, C. (2007). WCOT: A realistic lifetime metric for the performance evaluation of wireless sensor networks. *IEEE 18th International Symposium on Personal, Indoor and Mobile Radio Communications, 2007. PIMRC 2007*, Athens, pp. 1–5.

Rappaport, T.S. (2001). *Wireless Communications: Principles and Practice*, 2nd ed. Upper Saddle River, N.J.: Prentice Hall.

Reznik, L., and Von Pless, G. (2008). Neural networks for cognitive sensor networks. *IEEE International Joint Conference on Neural Networks (IJCNN)*, Hong Kong, pp. 1235–1241.

Rossi, M., Bui, N., and Zorzi, M. (2009, March). Cost and collision minimizing forwarding schemes for wireless sensor networks: Design, analysis and experimental validation. *IEEE Transactions on Mobile Computing* 8(3), 322–337.

Safwat, A., Hassanein, H., and Mouftah, H. (2001). Energy-efficient infrastructure formation in MANETs. *26th Annual IEEE Conference on Local Computer Networks, 2001. Proceedings. LCN 2001*, Tampa, Fla., pp. 542–549.

Shah, R.C., and Rabaey, J.M. (2002). Energy aware routing for low energy ad hoc sensor networks. *IEEE Wireless Communications and Networking Conference, 2002. WCNC2002*, Orlando, pp. 350–355.

Shariatpanahi, P., and Aarabi, H. (2007). On reliable routing in wireless networks with diversity. *IFIP International Conference on Wireless and Optical Communications Networks, 2007. WOCN '07*, Singapore, pp. 1–5.

Shu, H., and Liang, Q. (2005). Wireless sensor network lifetime analysis using interval type-2 fuzzy logic systems. *The 14th IEEE International Conference on Fuzzy Systems, 2005. FUZZ '05*, Reno, Nev., pp. 19–24.

Verdone, R., Dardari, D., Mazzini, G., and Conti, A. (2008). *Wireless Sensor and Actuator Networks; Technologies, Analysis and Design*. Great Britain: Elsevier.

Vijay, G., Bdira, E., and Ibnkahla, M. (2010). Cognitive approaches in wireless sensor networks: A survey. *25th Biennial Symposium on Communications (QBSC)*, Kingston, Ont., pp. 177–180.

Wang, G., Wang, T., Jia, W., Guo, M., Chen, H.-H., and Guizani, M. (2007). Local update-based routing protocol in wireless sensor networks with mobile sinks. *IEEE International Conference on Communications, 2007. ICC '07*, Glasgow, pp. 3094–3099.

Wang, Q., and Yang, W. (2007). Energy consumption model for power management in wireless sensor networks. *4th Annual IEEE Communications Society Conference on Sensor, Mesh and Ad Hoc Communications and Networks, 2007. SECON '07*, San Diego, Calif., pp. 142–151.

6

Enabling Cognition through Weighted Cognitive Maps

6.1 Introduction

A variety of tools have been utilized in the design of cognitive networks: neural networks, genetic algorithms, game theory, and expert systems, among others [1]. However, generally speaking, most of these tools either require long training periods or their complexity increases significantly as the considered variables increase.

In this chapter we consider weighted cognitive maps (WCMs) as a tool to provide a parametrized representation of the conflicting processes of the system. WCMs were first proposed in [13] and [14] as efficient tools to implement cognition in wireless sensor networks. In WCMs, each process, environment variable, or end-to-end goal is simply represented as a concept in the system, and the edges of the map connect concepts that are causally related. (An overview of WCMs is given in Section 6.3.) This means that the WCM deals with end-to-end goals or constraints as simple concepts in the system, considerably reducing the complexity. In addition, system interactions are represented by simple mathematical operations, thus avoiding the long processing times that may be associated with optimization problems. WCMs are also able to consider conflicting interactions within the network, requiring only information about the causal relationships of these processes.

Although WCMs were considered in [2] to design cognitive nodes, those nodes operated independently and the design did not consider interactions with other nodes. In addition, their design did not primarily target the needs of wireless sensor networks (WSNs). We attempt to utilize WCMs in order to design a cognitive engine for WSNs that can monitor network interactions and work to achieve its end-to-end goals while considering multiple constraints. Using WCMs this way will enable network designers to transcend the limitations of cross-layer design and introduce a degree of flexibility and adaptability in the network.

The remaining sections of this chapter are organized as follows. Section 6.2 reviews some recent related research efforts; Section 6.3 provides a brief introduction to WCMs; Section 6.4 explains the details of the cognitive WCM tool; Section 6.5 provides simulation results that illustrate the capabilities of the WCM tool, and finally Section 6.6 offers some concluding remarks.

6.2 Related Work

Several areas have been explored in WSN design: coverage and connectivity, routing, and topology management, among others. For example, the network management system proposed in [3] attempts to improve energy efficiency and network lifetime while considering routing and coverage constraints in a clustered hierarchy. An optimization problem is formulated, labeled OPT-ALL-RCC, that minimizes energy consumption and achieves load balancing while guaranteeing coverage and connectivity. OPT-ALL-RCC is shown to be NP-complete, and a heuristic scheme named TABU-RCC is proposed to achieve a compromise between efficient performance and processing time. A network management system known as energy-efficient m-coverage and n-connectivity routing (EECCR) was proposed in [4]. It considers the routing problem under multiple coverage and connectivity constraints. EECCR is divided into two main phases. The first phase sets up the routing paths by dividing the network into mutually exclusive scheduling sets, and switching on the sets that can guarantee m-coverage. Then routing paths are set up that achieve n-connectivity. The second phase in EECCR is the data transmission phase, where the set-up routing paths are utilized to relay the data to the sink node. In [5], a framework called topology-aware resource adaptation (TARA) was proposed with the primary goal of alleviating congestion in WSNs. The idea is to activate a larger number of nodes during periods of congestion in order to increase network resources and reduce congestion. Network topology and traffic patterns are considered in order to propose heuristics that can detect congestion, activate the correct number of nodes, and discover alternative routing paths that can relay packets away from the congested spots. TARA has the advantage of being able to operate efficiently in dynamic environments.

Even though WSN research is highly active, most of these efforts consider only specific issues, and limited efforts exist that are comprehensive in considering multiple issues that affect network efficiency. Optimization problems or heuristics are not easily extendable to consider multiple goals and constraints. Cognitive networking attempts to address these problems through the processes mentioned earlier.

In [1], Thomas et al. proposed a cognitive framework that is divided into three main layers. The highest layer is the requirements layer, which specifies the end-to-end goals of the network. The second layer is the processes layer, which contains all cognitive elements and algorithms that produce decisions to achieve the network goals. The third layer is the software adaptable network layer, which executes the decisions of the higher layers and gathers all the necessary knowledge for the operation of the cognitive framework. The framework does not specify any particular tools to implement the aforementioned layers but leaves this decision to the users according to the network requirements.

Other efforts in cognitive networks include the knowledge plane (KP) [6], which gathers knowledge in order to enable the network to adapt to different conditions, and detect and prevent different problems. Tools that have been utilized in the design of cognitive networks have been mentioned before. It is interesting to point out that designing a cognitive network for WSNs is different from other networks, such as ad hoc networks. In ad hoc networks, nodes may run different applications, and a certain degree of selfishness can be tolerated and cognitive nodes can afford to operate independently. On the other hand, in WSNs, nodes typically run the same application and have the same goal. For this reason, considering interactions between nodes within the entire network is highly important in WSNs if cognition is to be achieved.

6.3 Fundamentals of WCM

A WCM (also called fuzzy cognitive map) is a graphical model used to represent dynamical systems through their underlying causal relationships [7]. Each node in the WCM is called a concept and represents a particular process or event in the system being modeled. Edges of the WCM connect concepts that are causally related. Each concept C_i is characterized by a number A_i that represents the value of the concept or its activation level in the real system. Concepts can take values either in the interval [0, 1] or [–1, 1]. If the allowed interval is [0, 1], then a concept can be inactive ($A_i = 0$) or active with various levels ($A_i = 1$ means that a concept is fully active). If the allowed interval is [–1, 1], then concepts can also be decreasing. Edge weights can take on any value from the interval [–1, 1]. Negative edge weights imply negative causality, positive edge weights imply positive causality, and zero edge weights imply the absence of a causal relationship between concepts. WCMs can be qualitative or quantitative. Qualitative WCMs only represent causal relationships between concepts, while quantitative WCMs can also represent different levels of granularity in concepts. One of the main advantages of WCMs lies in their inference capabilities. To illustrate, consider the WCM representing processes of a wireless node shown in Figure 6.1.

The WCM of Figure 6.1 models the relationships between six processes affecting a wireless node: transmit power, data rate, maximum number of retransmissions allowed before a packet is dropped, bit error rate (BER), throughput, and expected transmission time (ETT). The edge weights shown

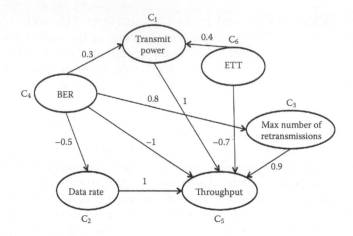

FIGURE 6.1
WCM representing processes of a wireless node.

represent the strength of causality between concepts. The WCM can be represented in matrix form as

$$
W = \begin{array}{c|cccccc}
 & C1 & C2 & C3 & C4 & C5 & C6 \\
\hline
C1 & 0 & 0 & 0 & 0 & 1 & 0 \\
C2 & 0 & 0 & 0 & 0 & 1 & 0 \\
C3 & 0 & 0 & 0 & 0 & 0.9 & 0 \\
C4 & 0.3 & -0.5 & 0.8 & 0 & -1 & 0 \\
C5 & 0 & 0 & 0 & 0 & 0 & 0 \\
C6 & 0.4 & 0 & 0 & 0 & -0.7 & 0 \\
\end{array} \qquad (6.1)
$$

For a WCM with n concepts, its status at time t can be given by

$$At = A_1 t \, A_2 t \, A_3 t \ldots An t^t \qquad (6.2)$$

According to the inference capabilities of WCMs, the status at time $t + 1$ is obtained by applying a threshold function $f(x)$ to each element of the product vector $WA(t)$ as:

$$A(t + 1) = f(WA(t)) \qquad (6.3)$$

Typically, $f(x)$ can be one of three functions [8]:

- Binary function

$$f(x) = \begin{cases} 1, & x > 0 \\ 0, & x \leq 0 \end{cases} \tag{6.4}$$

where the activation levels of concepts can either be 0 or 1, and the WCM in this case is called a simple WCM.

- Trivalent function

$$f(x) = \begin{cases} 1, & x > 0 \\ 0, & x = 0 \\ -1, & x < 0 \end{cases} \tag{6.5}$$

where concepts can take values from the set $\{-1, 0, 1\}$, and the WCM is called a trivalent WCM.

- Hyperbolic tangent function

$$f(x) = \tanh(x) \tag{6.6}$$

where the activation levels of the concepts can take any value from $[-1, 1]$, and the WCM is called a continuous WCM.

The inference process of WCMs is initialized when a particular concept is triggered. The triggered concept influences other concepts according to W. If the system is left free to interact through a series of matrix multiplication processes, it will finally reach one of three conditions [8]:

- An equilibrium, where further multiplications do not change the status of the WCM
- A limit cycle, where the system keeps returning to a specific status after a number of multiplications
- A chaotic status, where every new iterations result in new states

Simple and trivalent WCMs have finite number of states (2^n for simple WCMs and 3^n for trivalent WCMs), and therefore can only reach equilibrium points or limit cycles. Continuous WCMs can represent the system with finer granularity but may risk chaotic behavior, if not carefully designed.

WCMs have significant advantages over other tools, such as Bayesian and neural networks. They allow for feedback loops, which are not present in Bayesian networks. Concepts in WCMs also represent events or processes from the real system. This is not available in neural networks, which can be seen as a black box that is trained to model a particular system and may not faithfully reproduce its characteristics. The simple inference properties also make WCMs attractive for systems that require low complexity, such as WSNs. These advantages collectively enable WCMs to be used as a simple yet powerful tool that can accurately represent any dynamic system.

It is also worth noting that WCMs have some disadvantages [9]. They rely on expert knowledge to design the system, which may be challenging, especially in quantitative WCMs. Also, there is no research on how to build a WCM with a global view of the network, where multiple WCMs may exist and have to work together to achieve certain end-to-end goals.

6.4 Designing WCMs to Achieve Cognition in WSNs

The design takes into consideration some of the aforementioned challenges for WCM. In order to avoid the need for extensive training of the map, we use WCMs to formulate a parametrized representation of the WSN [13] [14]. This representation is neither fully quantitative nor qualitative but can rather be regarded as an intelligent decision support system that can produce qualitative decisions to determine specific values of different system parameters. Therefore, we aim at achieving the advantages of a quantitative WCM while maintaining the simplicity of a qualitative WCM and avoiding the need for extensive training of the map.

In order to build a distributed system of interacting WCMs that can work together to achieve the network's end-to-end goals, we start by considering a clustered WSN hierarchy. In the design we assume that the sink node and cluster heads (CH) are intelligent nodes, where the WCM will be implemented, while the sensor nodes are regular nodes with no intelligent capabilities and are randomly distributed over the coverage area. The CHs produce decisions that will be executed by the regular sensor nodes within their clusters. Since the only node that can have a centralized view of the network is the sink node, it will be used to monitor global concepts such as network connectivity and coverage. Thus, all concepts will be implemented at CHs and the sink node will make sure those concepts that require global monitoring are considered from the view of the entire network. Figure 6.2 illustrates the system architecture.

As Figure 6.2 shows, the sink node lies in the center of the area to be covered, while the CHs are distributed in key geographic positions, such that the number of regular nodes managed by each CH is fairly even. The sensor nodes are deployed throughout the coverage area. Each node is associated with the CH responsible for monitoring its cluster. We assume that network nodes are synchronized. They periodically wake up, sense the required attributes, transmit

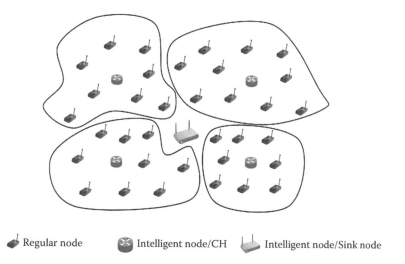

  Regular node   Intelligent node/CH   Intelligent node/Sink node

FIGURE 6.2
System architecture.

their measurements to the sink node, and go back to sleep mode. The fraction of this period where the node is awake is called the duty cycle.

Without loss of generalization, the WCM will consider transmit power, data rate, and duty cycle adaptation; coverage and connectivity; congestion; and routing. We define an end-to-end goal for the WCM that considers energy consumption as well as load balancing. These processes were chosen since they can have a significant impact on system performance. However, we stress that here the WCM system is not restricted to these processes. Other network designers may choose other processes that suit their requirements. Our aim is to illustrate the capabilities of WCMs and show how system processes, end-to-end goals, and variables can be translated into WCM concepts.

In the following, three WCMs are designed targeting a set of processes. These WCMs then are combined while achieving end-to-end goals [13] [14].

6.4.1 Designing a WCM for Transmit Power, Data Rate, and Duty Cycle Adaptation

In this section, a WCM is designed so that it uses transmit power, data rate, and duty cycle adaptation to improve energy efficiency, throughput, and link reliability, and ensure fairness among nodes. Before we show how a WCM can be designed for these processes, we first describe the protocols that will be represented by the WCM.

In Chapter 3, a protocol was proposed that adapts the duty cycle according to the data rate utilized (see e.g. [12]). The idea is that when nodes observe good channel conditions, they can transmit at higher data rates, thereby finishing their transmissions faster and going to sleep state sooner, and for longer periods [10]. This was shown to achieve considerable energy savings.

To illustrate, consider Figure 6.3, where P_{Tx} is the energy consumed during the active transmission period in watts, P_{idle} is the energy consumed during the idle period, T_d is the duration of the idle period, P_{sleep} is the energy consumed during the sleep period, and T_{Packet} is the duration of the entire cycle. As Figure 6.3 shows, when the utilized data rate is low (that is, the symbol duration is large), as in Figure 6.3(a), the transmission takes longer to finish than the case with high data rate (that is, in which the symbol duration is small), as in Figure 6.3(b). Thus the adaptive sleep period is longer in Figure 6.3(b), making it more energy efficient.

In this chapter we use this protocol and extend it to be implementable by the WCM tool.

In order to design a WCM for transmit power, data rate, and duty cycle adaptation, we first have to identify the environment variables that will be used to trigger the WCM to take action. We utilize the environment variables of packet loss ratio (PLR) and ETT. PLR was chosen because it accounts for channel conditions and interference, and can thus give a clear indication about the quality of the wireless link. ETT is the expected amount of time

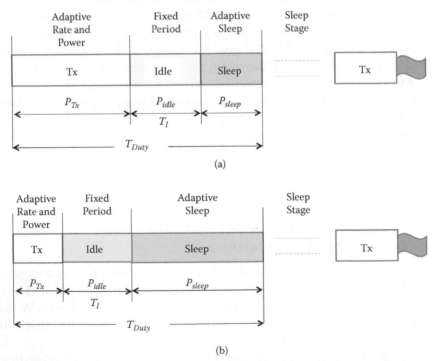

FIGURE 6.3
(a) Duty cycle with low data rate; (b) duty cycle with high data rate.

needed to successfully transmit a packet. It accounts for interference between links, number of retransmissions needed to successfully transmit a packet, and the data rate used at interfering links. Therefore, it can be used to improve throughput and ensure fairness between interfering links. For example, when the ETT of one link is significantly larger than that of another, this means that the node is not getting fair access to the wireless medium. By taking actions to keep ETT close among interfering links, fairness can be achieved.

Once there is a significant change in one of the environment variables, the WCM will be triggered to take action. The actions taken will be to adapt transmit power, data rate, or both. Adapting the duty cycle directly follows data rate adaptation. Particularly, when the data rate is increased, the duty cycle is decreased, and vice versa. We choose whether to adapt the transmit power or data rate according to which one results in lower energy consumption, since this is one of the primary objectives of a WSN. The energy consumed at a single node during one duty cycle can be expressed as

$$E = P_{Tx} \times \frac{L}{R} + P_{idle} \times T_d + P_{sleep}\left(T_{Packet} - T_d - \frac{L}{R}\right) \tag{6.7}$$

where L is the packet length, R is the data rate, and the remaining parameters were defined for Figure 6.3. When we need to choose whether to adapt transmit power or data rate, we need only compare the energy consumption resulting from such adaptations according to Equation (6.7). For example, when the PLR increases beyond a known threshold, the WCM needs to compare between increasing transmit power and decreasing data rate. After removing the constant parameters from Equation (6.7), this comparison can be expressed as

$$\frac{P_{Tx-Old}L}{R_{New}} - P_{sleep}\frac{L}{R_{New}} \geq \frac{P_{Tx-New}L}{R_{Old}} - P_{sleep}\frac{L}{R_{Old}} \tag{6.8}$$

where P_{Tx-Old} and R_{Old} are the transmit power and data rate values before adaptation takes place, while P_{Tx-New} is the value of transmit power if it will be increased and R_{New} is the value of data rate if it will be decreased. If the comparison in Equation (6.8) is "True," this means that decreasing the data rate will result in higher energy consumption than increasing transmit power, and thus the transmit power will be increased. If it is "False," then the data rate will be decreased instead. The WCM for transmit power, data rate, and duty cycle adaptation is shown in Figure 6.4, with concepts labeled C_1–C_5.

We use the concept of conditional edge weights in order to represent the causal relationships in the adaptation protocols. The causal relationships are used to formulate the matrix W, as in Equation (6.1). Thus when the PLR crosses a high or low threshold, a comparison according to Equation (6.8) is performed, and edges $W_{(1,2)}$ and $W_{(1,3)}$ are activated according to the following rule:

If PLR $< PLR_L$ or PLR $> PLR_H$ and Equation (6.8) is

$$\begin{cases} True, & W(1,2) = 1 \text{ and } W(1,3) = 0 \\ False, & W(1,2) = 0 \text{ and } W(1,3) = -1 \end{cases}$$

If PLR $> 2PLR$, then

$$W(1,2) = 1 \text{ and } W(1,3) = -1 \qquad (6.9)$$

where PLR_H and PLR_L are the values chosen for high and low PLR thresholds, respectively.

The rules in Expression (6.9) simply implement the algorithm detailed earlier (Eq. 6.8). However, we add a precaution that if the PLR is higher than $2PLR_H$, then this indicates a sharp deterioration in link quality, which has to be addressed by increasing transmit power and decreasing data rate simultaneously. The value $2PLR_H$ was chosen as it is not low enough to trigger frequent changes in transmit power and data rate, and not high enough to ignore a significant deterioration in link quality. Note that the WCM is triggered to take action if the PLR changes considerably or if the ETT of one node is significantly larger than that of an interfering node. As Figure 6.4 indicates, unfair access to the wireless medium is addressed by increasing transmit power.

After the weight matrix W is formulated, the CH formulates an array $A(t)$ similar to the one in Equation (6.2), reflecting the current activation levels of the concepts in Figure 6.4. A multiplication according to Equation (6.3) is performed to determine $A(t + 1)$, which determines the actions to be taken according to the new values of the concepts. It is important to stress that the result of this multiplication process will not specify the exact level of

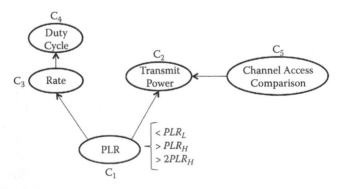

FIGURE 6.4
WCM of transmit power, data rate, and duty cycle adaptation.

transmit power or data rate to be used but only if they *should* be increased or decreased.

6.4.2 Designing a WCM to Guarantee Connectivity and Coverage

Two of the primary constraints of any WSN are to ensure that the network is always connected and that every point in the area is covered by at least k sensors. Typical WSNs have redundant nodes in order to extend their lifetime. Every node in the network wakes up in every duty cycle with probability p. This value should ensure that every point in the area is covered and that every node can find a routing path to the sink node. The value p is lower as the number of redundant nodes increases and should increase to 1 as more nodes die throughout the network's lifetime.

Here the objective of the WCM would be to adapt p in every duty cycle in order to ensure connectivity and coverage. In order to perform this task, we use the theorem derived in [11], which states that, for a sensing range r_s and communication range r_c at every node:

When $\alpha = r_s/r_c \leq 1$, the area *(D)* is almost surely (A.S.) connected-k-covered if, for some growing function $\varphi(np)$, p and r_s satisfy

$$np\pi r_s^2 \geq \log(np) + k \log\log(np) + \varphi(np) \qquad (6.10)$$

where n is the number of nodes in the network. The expression "connected-k-covered" means that every point in the network is covered by at least k sensors and can find a path to the sink node. A.S. connected-k-covered means that as n increases to infinity, the probability of connected-k-coverage goes to 1.

Note that as nodes die in the network, n will decrease accordingly. Thus the WCM will attempt to find p that satisfies Inequality (6.10) in every duty cycle. However, before this theorem can be used, we need to find an appropriate function $\varphi(np)$ that would ensure high probability of connected-k-coverage when Inequality (6.10) is satisfied (and checked through simulations).

In order to determine this function, we perform an experiment that is divided into two parts. In the first part, we perform computer simulations whereby we fix $p = 0.1$, $k = 1$, $r_s = 100$ m, and $r_c = 200$ m. The area to be covered is a square of size 500 m × 500 m. The number of nodes n is varied from 400 to 2000 nodes. For every value of n, we perform 1000 simulation runs, where every run consists of a new random deployment of nodes. In every run we check if the current deployment of nodes achieves connectivity and coverage. After 1000 runs, we calculate the probability that the network is connected and covered using this value of n. The simulation results are plotted in Figure 6.5.

In the second part of this experiment we use Inequality (6.10) with the same values of p, k, r_s, and r_c that were used in the simulations. We use a slowly growing function $\varphi(np) = (\log\log(np))^a$ and vary a from 4 to 5.5. For

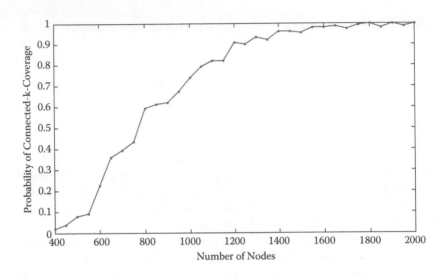

FIGURE 6.5
Probability of connectivity and coverage.

every value of *a*, we evaluate the minimum number of nodes *n* that would satisfy Inequality (6.10). Then we use the simulation results from Figure 6.5 to check the probability of connectivity and coverage using this value of *n*. The results are shown in Table 6.1.

TABLE 6.1

Values of *a* and Their Corresponding Probability of Connected-k-Coverage

Value of *a*	Minimum *n* to Satisfy Inequality (6.10)	Probability of Connected-k-Coverage (Obtained through Simulations)
4.0	876	0.60
4.1	904	0.62
4.2	934	0.65
4.3	966	0.68
4.4	1002	0.74
4.5	1041	0.78
4.6	1084	0.8
4.7	1131	0.81
4.8	1184	0.88
4.9	1242	0.89
5.0	1307	0.93
5.1	1379	0.95
5.2	1460	0.99
5.3	1550	0.99
5.4	1652	0.99
5.5	1767	0.99

As Table 6.1 shows, with $a = 4.0$, the minimum value of n that satisfies Inequality (6.10) is 876 nodes. However, the simulation results specify that the probability of connected-k-coverage is only 0.6 for this value of n. Therefore, $a = 4.0$ does not guarantee connectivity and coverage. On the other hand, with $a = 5.2$, the minimum value of n that satisfies Inequality (6.10) is 1460 nodes. At this value of n, the simulation results specify that the network is connected-k-covered with probability 0.99. Thus, this value of a provides a better guarantee for connectivity and coverage.

Given this result, we use the following function

$$\varphi(np) = \left(\mathrm{loglog}(np)\right)^a, \quad a = 5.2$$

After determining the appropriate $\varphi(np)$ to be used, we can implement this algorithm in the WCM. The concepts involved and the edges associated with them are shown in Figure 6.6. Each CH monitors the status of the nodes in its cluster and finds out the number of live nodes after detecting any failures due to battery expiration or any other conditions. If a failure is detected, then concept C_6 in Figure 6.6 is activated, prompting the CH to calculate the minimum p that would satisfy Inequality (6.10) for the new value of n. Once a new p is found, then the routing protocol will need to be invoked to find a new configuration for the new subset of nodes.

6.4.3 Designing a WCM for Congestion Control

Congestion typically occurs when a particular area in the network becomes exposed to an amount of traffic larger than the nodes can handle. This can cause queue build-ups and significant transmission delays. The traditional way of dealing with congestion is to instruct the source node to reduce its loading rate, which is the number of bits inserted in the transmission queue per second. Note that this is different from the node's data rate, which

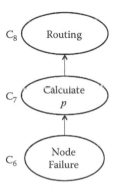

FIGURE 6.6
WCM to guarantee connectivity and coverage.

corresponds to the rate at which bits are transmitted from the transmission queue. Although reducing the source loading rate may reduce congestion, it has a significant flaw. From the point of view of the quality of service (QoS) observed by the user, this approach reduces the resources given to the user at the specific point when he or she is requesting more resources [5]. In WSNs, there is a possibility for another solution to this problem, which is to increase the amount of resources available in the network. This approach was explored in [5], where topology-aware resource adaptation (TARA) was proposed. In TARA, sleeping nodes are instructed to wake up around areas where congestion is occurring, and data is routed through these nodes to alleviate the problem. This approach may not always work over the entire network lifetime if nodes die and there are no more redundant nodes to wake up. It also requires significant knowledge of the network topology.

However, redundant nodes are not the only way of increasing network resources. Since we are already using duty cycle adaptation, as previously explained, we can use this feature to keep the nodes awake for longer periods of time during periods of congestion in order to dispense larger volumes of traffic. Route maintenance will still be needed to disperse paths that are causing congestion, but without the need for significant topology knowledge to know exactly which nodes need to be woken up. Using adaptive duty cycle will also work throughout the lifetime of the network. Examples of route maintenance operations to alleviate congestion are shown in Figure 6.7. In Figure 6.7(a), there is congestion around nodes A and B, since they are within interference range of each other and a large amount of traffic is flowing through them. In the system, the nodes will be instructed to extend their duty cycles to accommodate the extra traffic, and some of the traffic will be routed through node C. A similar process is done in Figure 6.7(b).

In order to implement this algorithm in a WCM, we first have to identify the appropriate parameters to detect congestion. We utilize the parameters of buffer capacity and channel utilization since they can be easily measured in real networks. Channel utilization is simply the fraction of time that a node detects the channel to be busy during a predefined interval, while buffer capacity is the number of remaining slots available in the buffer. Both parameters are

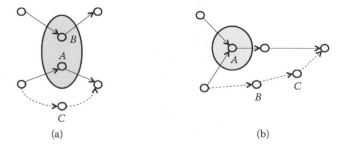

(a)   (b)

FIGURE 6.7
Examples of route maintenance operations.

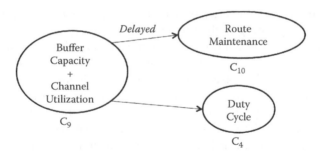

FIGURE 6.8
WCM algorithm for congestion control.

needed since only one of them may not give a clear indication about congestion. The corresponding WCM implementation is shown in Figure 6.8.

Once buffer capacity drops below a specific threshold *and* channel utilization is above a specific threshold, concept C_9 will be activated. This will prompt an increase in duty cycle and trigger the route maintenance operation. Note that route maintenance is delayed by one iteration, since it has to follow the increase in duty cycle.

6.4.4 End-to-End Goal and the Overall WCM

The choice of end-to-end goal is of primary importance to the performance of the system. All concepts will interact in order to achieve this goal. The network efficiency is chosen as the end-to-end goal of the WCM. We define network efficiency as the traffic load divided by energy consumption. The reason behind this choice is to give the network user the flexibility to request higher volumes of data throughout the lifetime of the network. We realize that higher volumes of data will ultimately require higher energy consumption, which has to be considered. Choosing energy consumption only or network lifetime only may cause the network to deny the user higher volumes of data in order to achieve the end-to-end goal. Therefore, a specific threshold for network efficiency will be calculated beforehand by the user, and the WCM will ensure that this threshold is not crossed. Each CH will monitor the network efficiency in its cluster, and the sink node will monitor the efficiency of the network.

Energy consumption in the system is the amount of battery declination within a given window divided by the residual battery power. Thus, every node maintains a moving time window and calculates how much the battery declines within this window. It then transmits this value along with its remaining battery power in every packet. Thus the energy consumption of node *i* can be given by

$$Eng\_consump_i = \frac{E_{start\_window} - E_{end\_window}}{E_\eta} \tag{6.11}$$

where E_{ri} is the residual energy of node i, E_{start_window} is the remaining energy of node i at the beginning of the time window, and E_{end_window} is the remaining energy at the end of the window. Equation (6.11) considers remaining battery power as well as load balancing. If the remaining battery power of a node is low, its efficiency will decrease, prompting the WCM to take action. If its energy consumption increases, say from being involved in excess traffic, its efficiency also decreases, again prompting the WCM to take action. This ensures that nodes are not overused and will be avoided if their battery power is low.

Moreover we define traffic load as the amount of traffic dispensed by any node within a given time window. Thus, the traffic load of node i can be expressed as

$$\text{Load}_i = \text{Number of bits dispensed within window of length } t \,/$$
$$\text{Total number of bits input to } i \text{ within } t$$

$$= \frac{R_i \times t_{Transmission}}{S_i \times t} \tag{6.12}$$

where R_i is the data rate of node i, t is the length of the window, $t_{Transmission}$ is the fraction of the time window that node i is transmitting, and S_i is the source loading rate of node i. Therefore the efficiency of a cluster with number of nodes G is given by

$$\text{Efficiency} = \sum_{i=1}^{G} \frac{\text{Load}_i}{\text{Eng}\_\text{consump}_i} \tag{6.13}$$

From Equation (6.11), Equation (6.12), and Equation (6.13), the WCM will take action if the rate of battery declination increases, if the data rate decreases, or if the source loading rate increases. The action to be taken depends on the status of the nodes. If the PLR is low enough, then nodes can be instructed to decrease transmission power or increase data rate by performing the comparison that was specified in Equation (6.8). If the PLR is not low enough, then the routing module should be invoked to find routing configurations that avoid nodes with low efficiency as much as possible. The final version of the WCM is shown in Figure 6.9.

As Figure 6.9 shows, all the dashed edges are those going in and out of the end-to-end goal of efficiency. Figure 6.9 also shows a concept that is only activated if transmit power and data rate adaptation can no longer repair a specific link. The link is deemed failed and has to be maintained. Similarly, if adaptation of transmit power fails to address the issue of ETT imbalance, then the route maintenance module is invoked to repair part of the route where the problem is occurring. Note that the difference between the route maintenance module and the routing module is that the route maintenance

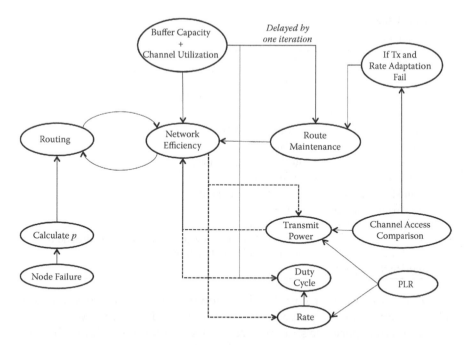

FIGURE 6.9
Overall WCM of the cognitive network.

module only repairs part of the route, while the routing module is invoked to find complete new paths. For this reason, the routing module is invoked when a node's efficiency decreases, for example, since this node will need to be removed from some of the routing paths in which it may be participating, thus requiring extensive modifications to existing routing paths. On the other hand, link failures or congestion only require that packets be routed away from the problematic areas, thus only a few links need to be replaced; in this case, the routing maintenance module is invoked.

The WCM shown in Figure 6.9 is implemented at the CH and the sink node, which will use it to make the necessary decisions for determining the parameters to be used for node operations. The WCM considers several parameters and issues of importance to network performance. Also the decisions taken by the WCM are adaptive and depend on the particular situation that caused the WCM to react.

6.5 Simulation Results

In this section the performance of the cognitive tool is evaluated through computer simulations. The propagation model we use is lognormal shadowing.

TABLE 6.2

Simulation Parameters

Parameter	Value
Grid size	500×500 m
Number of nodes	169, 256, 400, 625, and 900
Transmit power levels	[–12:6:36] dBm
Data rates	6 Mbps $\rightarrow$ 54 Mbps
Cycle period	100 ms
Duty cycle	$[0.25, 0.5, 0.75, 1] \times 100$ ms
Packet size	1000 bytes
Source loading rates	$[0.2, 0.4, 0.6, 0.8, 1] \times 10^4$
Initial battery power	5 Ahr
Power consumption	Rx: 26 mAhr
	Tx: 26 mAhr + (Tx power $\times$ packet size $\times$ bit duration)
	Sleep: 0.3 µAhr $\times$ sleep time
Initial buffer capacity	100 packets

In addition, we consider correlated shadowing in order to ensure a comprehensive and realistic propagation environment. Thus, links within close proximity will experience similar or correlated shadowing values. The main simulation parameters are shown in Table 6.2.

6.5.1 Evaluation Using Uniform Random Topology

The performance of the system is evaluated in this section using a uniform random topology, similar to the one that was previously explained in Figure 6.2. The performance of the system is compared to the network management system known as TABU-RCC, which was reviewed in Section 6.2. This is because the end-to-end goal of TABU-RCC is to maximize network lifetime in the presence of coverage and connectivity constraints, which is similar to the goals and constraints considered by the WCM system. We also compare the WCM system to a regular network with no management or adaptation protocols, in order to provide a baseline for our comparisons. We consider the metrics of network lifetime, throughput, PLR, and the amount of information that the network is able to dissipate before it reaches its lifetime. Note that the amount of information dissipated before the network expires reflects the efficiency of the system in utilizing the available resources in transmitting packets. We define network lifetime as the time spanning from the start of network operation until the time when the remaining nodes can no longer guarantee full coverage and connectivity of the network.

In all the simulations conducted, we divide the time into cycles, or rounds. In every cycle, nodes wake up according to probability *p*. In WCM and TABU-RCC, this value changes throughout network lifetime to maximize energy efficiency. Nodes will transmit packets according to the source loading rate, and according to the routing path, transmit power, and data rate specified by the WCM. At some point during the cycle, the node will go back to sleep mode according to the duty cycle, also specified by the WCM, and the process repeats itself. We keep the source loading rate, initial buffer capacity, initial battery power, packet size, and cycle period constant in all simulations in order to ensure fair comparisons between different protocols.

In the first experiment, we evaluate the lifetime of the system under different number of nodes, compared to TABU-RCC and a "regular" network with no management or adaptation protocols. The simulation results are displayed in Figure 6.10, which shows that the WCM tool achieves longer lifetime than the other protocols, especially in larger networks. In smaller networks, WCM is forced to switch most of the nodes in the network to active mode in every duty cycle in order to guarantee connectivity and coverage. Thus network management does not have as big an impact in smaller networks as in larger ones. However, adaptation processes, such as transmit power and data rate adaptation, enable WCM to achieve a slightly longer lifetime than TABU-RCC in smaller networks. In larger networks, WCM achieves good lifetime results, clearly showing the efficiency of WCM as a network management system. Note that the regular network with no management or adaptation achieves poor performance results, since it has no capabilities to utilize the redundant nodes available. Thus all nodes are switched on in every duty cycle, regardless of the number of nodes in the network, which is why lifetime here is constant for all node sizes.

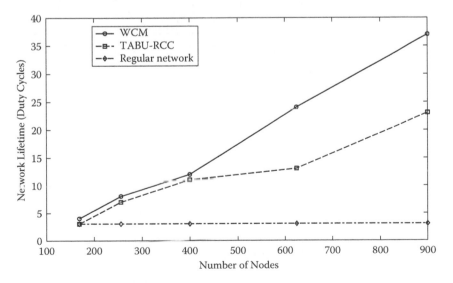

FIGURE 6.10
Network lifetime of WCM compared to other systems.

In the next experiment, the performance of the WCM system is evaluated in terms of throughput and PLR, compared to TABU-RCC and the regular network. The simulation results are shown in Figure 6.11 and Figure 6.12.

As Figure 6.11 shows, WCM achieves the highest throughput results, especially as the size of the network increases. In smaller networks the impact of

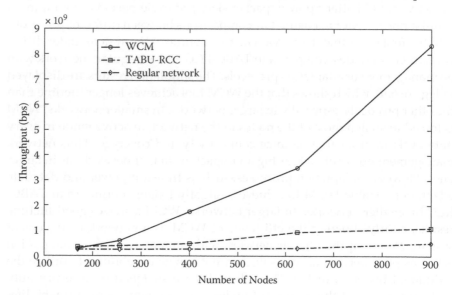

FIGURE 6.11
Throughput of WCM compared to other systems.

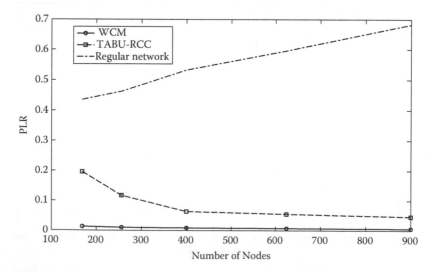

FIGURE 6.12
PLR of WCM compared to other systems.

the WCM system is minimized. As the size of the network increases, transmit power and data rate adaptations have a larger impact, especially as the network topology changes throughout its lifetime, due to the deaths of nodes. This illustrates the ability of WCM to adapt to changing network conditions.

Figure 6.12 shows that the WCM system also achieves the lowest PLR results, due to the efficiency of transmit power and data rate adaptations. TABU-RCC also achieves low PLR results, since it is capable of switching on the minimum number of nodes in every duty cycle, thereby reducing interference. The regular network with no adaptations or management suffers from high PLR, since all nodes are switched on in every duty cycle, which causes high interference and collisions. We can also see that the PLR of the regular network increases as the number of nodes increase, since increasing the size of the network causes more interference and more collisions.

In the next experiment, we evaluate the capability of the WCM system to utilize resources in disseminating information. This is done by counting the total number of packets transmitted by all nodes in the network throughout its lifetime. The numbers are compared with those of TABU-RCC and the regular network.

Figure 6.13 shows that WCM is able to transmit a significantly larger number of packets during network lifetime, especially in larger networks. The results from this experiment combined with Figure 6.11 and Figure 6.12 illustrate how WCM can utilize the available resources in maximizing network efficiency. We can also see from Figure 6.13 that the regular network transmits the largest number of packets in small networks, since all nodes are

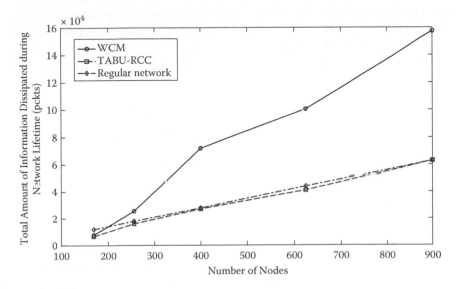

FIGURE 6.13
Capability of WCM to disseminate information compared to other systems.

switched on. TABU-RCC transmits a number of packets that are close to the regular network. However, we have seen from Figure 6.12 that the PLR in the case of the regular network is quite high, which means that a significant portion of the transmitted packets are lost. Thus, this type of network is not very efficient.

6.5.2 Evaluation Using Bottleneck Paths

In this section the performance of the WCM tool is evaluated under challenging network scenarios. The goal is to see if the WCM can adapt to different network topologies. We use a topology where transmissions are forced to go through a limited set of nodes, thus forming a bottleneck area in the network.

As Figure 6.14 shows, all transmissions are forced to go through a limited set of paths. In this section the performance of the WCM system is also evaluated using the metrics of network lifetime, throughput, PLR, and the total number of packets transmitted during the lifetime of the network. We also keep the source loading rate, initial buffer capacity, initial battery power, packet size, and cycle period constant in all simulations in order to ensure fair comparisons between different protocols. Figure 6.15 shows the network lifetime results of WCM compared to TABU-RCC and the regular network with no adaptations.

As Figure 6.15 shows, WCM achieves good lifetime results, especially in larger networks. Note that in small networks, TABU-RCC and regular networks did not work, as they could not adapt their parameters to guarantee network connectivity. Also their lifetime is the same for any network size. This is because the lifetime of these systems ultimately depends on the lifetime of the bottleneck nodes, and the number of bottleneck nodes does not change as we increase network size. However, using WCM, transmit power and data rate can be adapted such that not all bottleneck nodes are used in

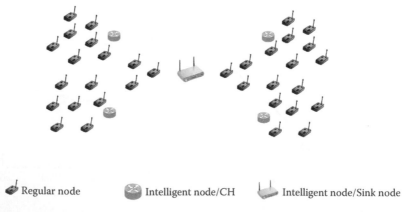

　🐞 Regular node　　　🔲 Intelligent node/CH　　　📶 Intelligent node/Sink node

FIGURE 6.14
Network topology with bottleneck path.

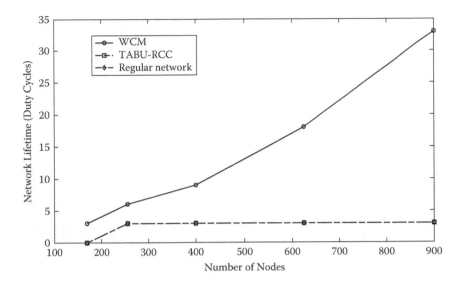

FIGURE 6.15
Network lifetime of WCM in bottleneck path.

every transmission, thus providing better load balancing, which increases network lifetime.

In the next experiment, we evaluate the performance of WCM in terms of throughput and PLR. The simulation results are shown in Figure 6.16 and

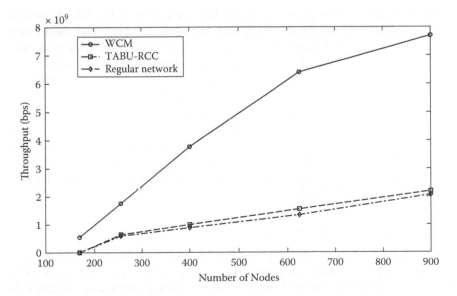

FIGURE 6.16
Throughput of WCM in bottleneck path.

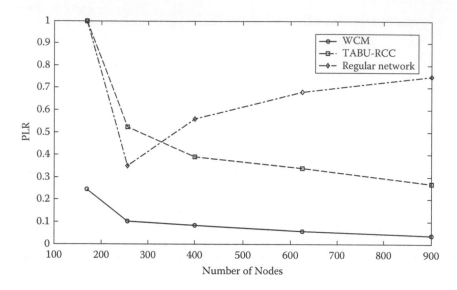

FIGURE 6.17
PLR of WCM in bottleneck path.

Figure 6.17, which show that WCM achieves good performance results in terms of throughput and PLR, and the performance improves in larger networks. Figure 6.17 also shows that the PLR of the regular network increases as the number of nodes in the network increases. This confirms the results in Figure 6.12.

In the final experiment, we evaluate the capability of the WCM packets to disseminate information throughout network lifetime. The simulation results are shown in Figure 6.18. WCM is able to utilize network resources more efficiently, and thus is able to dissipate a larger number of packets during network lifetime. In smaller networks the numbers are close because the available resources are limited. Figure 6.18 also shows that the capability of TABU-RCC to disseminate information is close to that of a regular network with no adaptations or management. This is because the bottleneck path has limited capacity, and therefore it becomes the dominant factor in determining how many packets can be transmitted throughout network lifetime. WCM overcomes this drawback because it has the ability to adjust parameters, providing better utilization of nodes.

6.5.3 Complexity of the System

The complexity of a system can be classified into two main parts: computational complexity and communication overhead. Computational complexity can be defined in several ways, which makes it sometimes difficult to quantify. In this chapter we refer to computational complexity as the number of steps needed to execute the protocol, which is directly proportional to

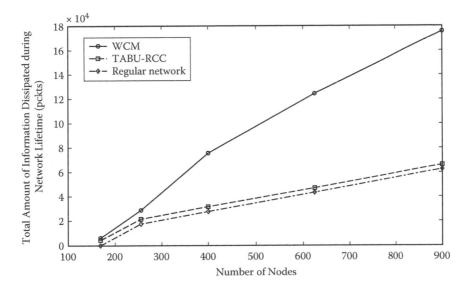

FIGURE 6.18
Capability of WCM to dissipate information in bottleneck path.

the processing time. On the other hand, communication overhead typically refers to the extra control packets needed for the protocol to operate.

The execution of the protocol requires simply the WCM matrix multiplication operation. The activated concepts determine the input array, and the status of the nodes determines the WCM matrix. Thus the protocol does not require extensive search operations or loops that may be required by optimization problems. This makes the processing time of the system relatively short, which also means that it can react quickly to network conditions before they change. It is important also to note that the computational complexity of the WCM system is directly proportional to the number of concepts included. This gives the network designer a degree of flexibility in choosing the required level of complexity. For example, if traffic in the network is expected to be light, one can remove concepts that consider congestion to make execution faster. The network designer has to ensure that the performance gain from the concepts included outweighs the incurred complexity.

The system also requires a small increase in communication overhead. The information required for protocol execution, such as PLR levels, battery consumption, and channel utilization, can simply be piggybacked on transmitted data packets. Special packets may be occasionally needed if some nodes need instant action from the CH. Note that in WSN, nodes typically transmit sensed information periodically, which means that data packets will be regularly available for piggybacking information.

6.6 Conclusion

In this chapter we presented a cognitive network management protocol for WSNs based on WCM. The WCM system is able to perform efficient reasoning while considering multiple objectives and constraints. By maintaining an overview of all network elements, the WCM is able to ensure they all operate coherently. The WCM constantly monitors the required QoS levels specified by the user and takes fast and efficient actions whenever those levels are violated. This is achieved while avoiding the high complexity typically required by optimization problems.

To evaluate the performance of the system, extensive computer simulations were conducted, and the WCM was compared against other well-known protocols. The WCM was shown to be highly adaptive and dynamic in detecting changes in the network and reacting quickly to them. For these reasons, the WCM was shown to outperform other protocols in metrics of lifetime, throughput, and PLR.

References

1. R. Thomas, D. Friend, L. DaSilva, and A. MacKenzie, "Cognitive networks," Chapter 2 in H. Arslan (Ed.) *Cognitive Radio, Software Defined Radio, and Adaptive Wireless Systems,* Springer, 2007, pp. 17–41.
2. C. Facchini and F. Granelli, "Towards a model for quantitative reasoning in cognitive nodes," *IEEE Global Telecommunications Conference (GLOBECOM),* December 2009, pp. 1–6.
3. A. Chamam and S. Pierre, "On the planning of wireless sensor networks: Energy-efficient clustering under the joint routing and coverage constraint," *IEEE Transactions on Mobile Computing* 8, no. 8, August 2009, pp. 1077–1086.
4. Y. Jin, L. Wang, J.-Y. Jo, Y. Kim, M. Yang, and Y. Jiang, "EECCR: An energy-efficient *m*-coverage and *n*-connectivity routing algorithm under border effects in heterogeneous sensor networks," *IEEE Transactions on Vehicular Technology* 58, no. 3, March 2009, pp. 1429–1442.
5. J. Kang, Y. Zhang, and B. Nath, "TARA: Topology-aware resource adaptation to alleviate congestion in sensor networks," *IEEE Transactions on Parallel Distributed Systems* 18, no. 7, July 2007, pp. 919–931.
6. G. Vijay, E. Bdira, and M. Ibnkahla, "Cognition in wireless sensor networks: A perspective," *IEEE Sensors Journal* 11, no. 3, March 2011, pp. 582–592.
7. J. Dickerson and B. Kosko, "Virtual worlds as fuzzy cognitive maps," *IEEE Virtual Reality Annual International Symposium,* September 1993, pp. 471–477.
8. A. Tsadiras, "Comparing the inference capabilities of binary, trivalent and sigmoid fuzzy cognitive maps," *Journal on Information Sciences* 178, no. 20, October 2008, pp. 3880–3894.

9. X. Luo and X. Zhang, "Guided gamed-based learning using fuzzy cognitive maps," *IEEE Transactions on Learning Technologies* 3, no. 4, October 2010, pp. 344–357.

10. X. Zhao, E. Bdira, and M. Ibnkahla, "Joint adaptive modulation and adaptive MAC protocols for wireless sensor networks," submitted to *International Journal of Distributed Sensor Networks,* December 2011.

11. G. Yang and D. Qiao, "Critical conditions for connected-k-coverage in sensor networks," *IEEE Communications Letters* 12, no. 9, September 2008, pp. 651–653.

12. K. Ramachandran, R. Kokku, H. Zhang, and M. Gruteser, "Symphony: Synchronous two-phase rate and power control in 802.11 WLANs," *IEEE/ACM Transactions on Networking* 18, no. 4, August 2010, pp. 1289–1302.

13. A. El-Mougy, *Weighted Cognitive Maps for Cognitive Wireless Sensor Networks: Design and Analytical Analysis,* PhD diss., Queen's University, 2012.

14. A. El-Mougy and M. Ibnkahla, "Cognitive WSN design using weighted cognitive maps," *IEEE Transactions on Wireless Communications* (submitted for publication), 2012.

7

Hardware Architecture for GPS/INS-Enabled Wireless Sensor Networks

7.1 Introduction

Hardware implementation of cognitive wireless sensor networks (WSNs) is still at an early development stage, and very few works have addressed this issue. In [24] and [25], the authors have implemented two cognitive architectures where cognitive nodes play an important role. The first architecture involves a single cognitive node equipped with high computational and power capabilities. Its role is to collect environment, user, and network information and make power management and scheduling decisions. The second architecture involves one cognitive node per cluster or per group of clusters. Field test results show significant lifetime improvements compared to noncognitive networks. Also, the idea of having specialized cognitive sensors is fully justified since the cognitive sensor needs to have specific features that are not available in a "regular" cluster head, for example.

This chapter focuses on hardware implementation of wireless sensor networks where location information is of paramount importance. In many WSN applications, distributed sensor nodes collect data at different locations, and the location information of each node is often required. Moreover, the location information can also be used by the telecommunications system itself, such as routing protocols. A typical example is a location-aided routing (LAR) protocol, which utilizes location information to decrease overhead of route discovery [1]. Many other routing protocols such as geographic random forwarding (GeRAF) are also based on geographical location of nodes [2] [3] [4]. Furthermore, cognitive approaches often require position information, such as cognitive diversity routing (Chapter 5).

The global positioning system (GPS) is usually used to identify the spatial coordinates of a sensor node in a WSN. Generally, GPS requires direct line-of-sight (LOS) signals from at least four satellites to estimate the receiver's coordinates. A stand-alone GPS may often suffer from signal blockages in degraded signal environments, such as indoor environments, urban canyons, and so forth. The integration of a strap-down inertial navigation system (SINS) with the GPS has been extensively studied and deployed in different applications. In these integrated systems, SINS provides position, velocity, and orientation information at very high rates (usually above 50 Hz) and provides accurate outputs in a short term. However, its performance deteriorates quickly with time. Therefore, GPS and SINS are complementary and each system compensates for the other's drawbacks. It is important to integrate GPS with SINS for continuous reliable navigation information.

Standard inertial navigation systems (INSs) use high-cost accelerometers and gyroscopes to provide precise navigation information, which lessens their popularity in general users' end devices. With the advancement of micro-electromechanical system (MEMS) technology, low-cost MEMS inertial sensors provide a more affordable solution for GPS/INS integrated navigation systems [5] [6] [7] [8]. Although MEMS sensors make the navigation system less expensive, more compact, and more power efficient, the performance is relatively poor due to its high drifts, its vulnerability to temperature effects, and so on. As a result, INS errors accumulate rapidly in short time intervals unless there are updates from external navigation measurements. These errors can usually be reduced by high-quality integration algorithms such as a Kalman filter (KF) with regular updates, so that the accuracy can always be kept at an acceptable level. Kalman filters are widely used as a common data fusion algorithm in many SINS/GPS applications [7] [8] [9]. However, Kalman filters have some shortcomings. Since it is the optimal filter for modeled processes, predefined system dynamic models are required to make it work well. If the input data does not fit those models, the output will not be satisfactory [10]. The observability of some of the error states is another major problem. Moreover, the performance of Kalman filters may be significantly degraded if the sensor noise is high [11]. Although some alternative algorithms for INS/GPS integration (such as artificial neural networks) have been investigated and proved efficient in navigation applications, KF is still computationally efficient and particularly suitable for real-time applications. Therefore, KF is used here as the navigation algorithm to fuse the data outputs from GPS and inertial sensors.

In land navigation applications, the system is assumed to sometimes operate in weak signal environments; that is, short-term GPS outage occurs frequently. Some previous research has successfully shown that the integrated system was able to provide accurate and reliable navigation solutions during GPS outages. A NovAtel Black Diamond System, which combines a NovAtel OEM4 GPS receiver and a Honeywell HG1700 inertial measurement unit (IMU), has been tested and the system has been reported to work well during long GPS gaps. Furthermore, accuracy could be improved by using field calibration techniques, velocity matching alignment, and nonholonomic constraints [12].

Recently, many research efforts have focused on GPS/INS integration techniques implemented in real-time embedded systems. Here, MEMS-based

IMU plays an essential part to make the whole systems cost-effective and compact.

A micro-miniature inertial measurement unit (MIMU)/GPS integrated system has been proposed by [13]. This system is based on a PC/104-embedded microcomputer, and the MIMU module is composed of three MEMS gyroscopes and three MEMS accelerometers. Six inertial sensor signals are processed by a multiplexer-based analog-to-digital converter (ADC) on a customized data acquisition board. The micro-programming controlled direct memory access (MCDMA) technique is introduced to improve the real-time data calculation and achieve fast data exchange.

Qu et al. [14] have described a miniaturized/low-cost navigation system in which a digital signal processor (DSP) and a complex programmable logic device (CPLD) are used as a coprocessor for developing the hardware system. In this system design, CPLD, which has a fast calculation speed, is mainly used for matrix calculation, and the DSP is used to control and schedule the entire system.

The authors in [15] describe the development of a compact and low-power GPS/INS system based on a field-programmable gate array (FPGA) and a floating point DSP. In this paper, they used the FPGA to make an efficient interface for GPS data acquisition. An internal dual port random access memory (DPRAM) within the FPGA is used for asynchronous data transmission between the GPS and the DSP, which significantly reduced the GPS's processing overhead on the navigation processor. It adopted a 16 bit, 250 kHz analog-to-digital converter (ADC) (ADS8364, Texas Instruments [16]) for sampling all the analog signals from inertial sensors simultaneously, rather than using an ADC multiplexer to sample the channels at different time instants.

This chapter describes a low-power miniaturized navigation system design for sensor nodes in wireless sensor networks (WSNs), where cost, size, and energy consumption are focal concerns. The system was originally proposed in [23]. A 2D loosely coupled GPS/INS integrated approach is applied. Details of the development of this real-time system based on one single fixed-point DSP are presented. A single power supply of 4.5 V is used in the proposed system. To minimize the power consumption, this system uses a programmable interface controller (PIC) microcontroller to detect the motion/acceleration of the navigation board and to control the power supplies of various circuits and integrated circuits (ICs). The motion detection scheme takes advantage of the ultra-low-power wake-up feature of the micro controller unit (MCU) PIC16F886, which significantly decreases the power consumption of the board when it is stationary.

The remainder of this chapter is organized as follows. Section 7.2 presents the architecture of the system. Section 7.3 describes details of the software implementation. Experimental examples are provided in Section 7.4.

7.2 Hardware Implementation

The hardware architecture of the system is presented in Figure 7.1. The final navigation board layout is shown in Figure 7.2. The components of the system can be divided into five main blocks:

- Data acquisition components: inertial sensors, GPS, magnetic sensor
- Data processing unit: DSP
- Power management unit: microcontroller
- Wireless transceiver
- Power supply

Each of these parts is described in the following sections.

7.2.1 GPS and INS Data Acquisition

For the low-cost objective, the IMU on the navigation board consists of a reduced set of inertial sensors involving one single-axis ±80°/s yaw rate gyroscope ADIS16060 [17] and one dual-axis ±1.7 g accelerometer ADIS16003 [18] from Analog Devices Inc. Both the accelerometer and the gyroscope operate on a voltage supply of 3.3 V. The specifications are given in Section 7.4.

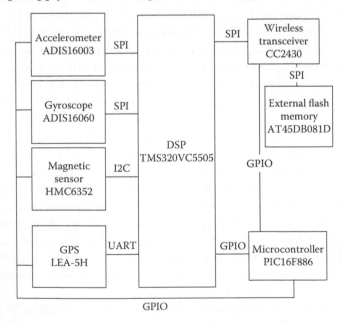

FIGURE 7.1
System architecture.

FIGURE 7.2
System's printed circuit boards (PCB).

A GPS receiver LEA-5H from μ-blox is used in the proposed integrated system. Its compact size, low power consumption, highly reliable positioning performance (2.5 m circular error probable (CEP)) and its SuperSense feature (high sensitivity and high jamming immunity [19] with 18 tracking channels) make it an outstanding choice among off-the-shelf commercial products.

The GPS module transmits data using the universal asynchronous receiver/transmitter (UART) interface and it uses National Marine Electronics Association (NMEA) data communication protocol to transmit position information. The serial data, in the form of NMEA sentences, are transmitted in ASCII code. The data acquisition process is performed on DSP, which parses the NMEA sentences; extracts useful information such as latitude, longitude, and time tag; and then converts that information into binary code.

7.2.2 Navigation Data Processing

The DSP used for this design is a 196-pin fixed-point TMS320VC5505, which is a member of Texas Instrument's (TI's) TMS320C5000 family. This DSP is well designed for low power applications, capable of operating at a maximum CPU frequency of 100 MHz when the core voltage is 1.3 V.

To improve the real-time performance and reduce the data acquisition overheads, FPGAs are used in many GPS/INS applications to assist the DSP in receiving navigation data. The integration of FPGA and DSP can significantly improve the calculation efficiency since the DSP can directly fetch the data instead of waiting for the low-speed serial input/output (I/O) operation. Although the real-time system may benefit from FPGA in high-sampling data-rate cases, a single-chip DSP can provide comparable performance when the sampling data rate is low. And most importantly, a single DSP makes the navigation system simpler and more compact, which meets the requirements of energy-sensitive applications and cost-sensitive WSN applications.

7.2.3 Power Management MCU

Power consumption is usually one of the dominating design criteria in wireless sensor network applications. To minimize the system's power, a motion detection scheme is used to optimize the power management. When the WSN node is stationary and its position is fixed, both GPS and inertial sensors will be turned off until it starts to move again.

PIC16F886 MCU is used to accomplish the motion detection design. It is featured with an ultra-low-power wake-up function that provides an efficient technique by periodically waking up the MCU from sleep. It charges a capacitor, enables a low voltage interrupt, and goes to sleep. When this capacitor slowly discharges below V_{IL}, a wake-up interrupt will wake up the controller again.

The design idea of the motion detection scheme is to keep the PIC MCU alternately going between sleeping and waking up to maximize the sleep time of the microcontroller so that the overall power consumption is minimized. Figure 7.3 shows the flow chart of the motion detection scheme. The microcontroller waits for the external interrupt for a certain period (that is, the detection period is set by a timer). If there is no external interrupt during the detection period, a timer interrupt will be generated and the microcontroller will go into ultra-low-power sleep mode. Otherwise, if the target shakes or a movement is detected, a counter will increase and the MCU will go into the motion reconfirming mode and wait for the next external interrupt. If the counter reaches the motion threshold, then the MCU will send a message to the DSP indicating that a constant motion is detected.

7.2.4 Wireless Radio Frequency Transceiver

The CC2430 is a system-on-chip solution from TI that contains a high-performance and low-power 8051 processor core and 2.4 GHz IEEE 802.15.4–compliant radio frequency (RF) transceiver. It is highly suited for systems that require ultra-low-power consumption. When the microcontroller is running at 32 MHz, the current consumption of the transmitter/receiver is typically as low as 27 mA. In addition, it has four flexible power modes with very short transaction time between low-power modes and active mode, which can effectively reduce the average power consumption in low duty-cycle systems.

To measure the communication range, a range test was conducted in an open field environment. In this test, the transmitter continuously sends data packets to the receiver. CC2430 has a built-in received signal strength indicator (RSSI), giving a digital value that can be read from one of its special function registers. The RSSI used in this test was the average RSSI of the last

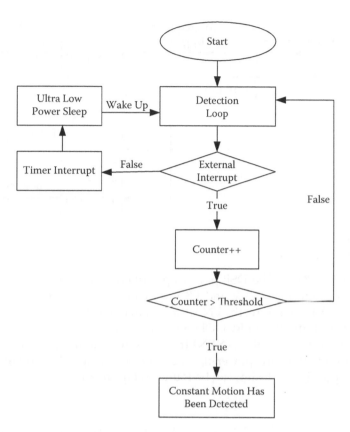

FIGURE 7.3
Motion detection flow chart.

32 received packets. The packet error rate (PER) and RSSI are the parameters that are used to determine the communication quality of this test.

Based on the range test, the effective range of the radio link between two CC2430 nodes is approximately 277 meters, when the transmit power is 19 dBm (with a device current consumption of 32.4 mA [20]), under line-of-sight (LOS) propagation.

7.2.5 Power Supply

The power supply set required for the navigation board is [1.3 V, 1.8 V, 3.3 V, 5 V]. The 3.3 V and 5 V are used by all the peripherals. The DSP requires 1.3 V and 1.8 V for core power and 3.3 V for the I/O pins. Three AA-size batteries (3 × 1.5 V) are used to provide the DC power, which is initially regulated at 5 V by a fixed-output boost converter TPS61032. This converter also provides interrupt signals required by the DSP. The 1.3 V, 1.8 V, and 3.3 V power supplies are provided by using three adjustable low-dropout voltage

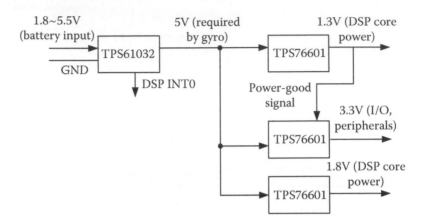

FIGURE 7.4
Power circuit design.

regulators TPS76601. The DSP power-up sequence specifically requires that the core-level supplies (1.3 V and 1.8 V) must power up before the I/O level supplies (3.3 V). Therefore, the ENABLE signal of the 5 V-to-3.3 V converter is connected with the POWERGOOD pin of the 5 V-to-1.3 V converter through a negative-positive-negative (NPN) transistor to guarantee a sufficient time delay between the core power-up and I/O power-up. More details of the power supply circuit design can be found in Figure 7.4

7.3 System Software Design

7.3.1 System Initialization

The system initialization routine includes DSP clock frequency configuration, parallel port configuration, and the initialization of its peripherals (see Figure 7.5). Various external interrupts and interrupt service routines (ISRs) are described as follows:

- Whenever a reset signal is generated at pin D6 of C5505, the DSP terminates execution and loads the program counter with the contents of the reset vector, which leads the program to return to the on-chip ROM bootloader. After completing the reset ISR, the program restarts the initialization function.

- When the battery voltage is lower than 1.8 V (making the voltage at low battery input (LBI) lower than 500 mV), the low-power detection circuit causes the low battery output (LBO) pin to generate a logic low signal that forces the program to jump into the BatteryPower_ISR.

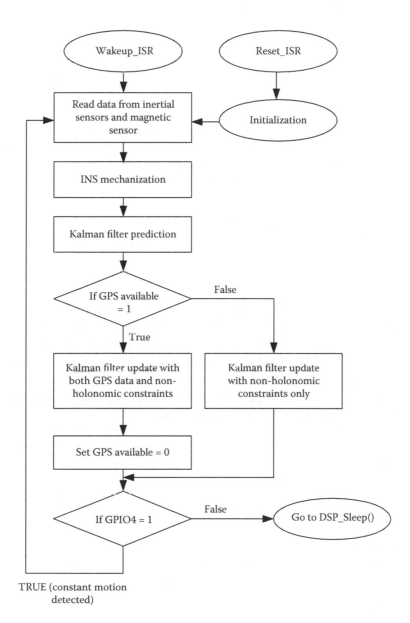

FIGURE 7.5
System software flow chart.

This ISR stops the program execution and the LED on pin M8 flashes as a low-battery-power warning.

- When over 70 bytes of serial data (approximately as long as a GPS-recommended minimum sentence (GPRMC) sentence with full information) are sent from the GPS to the DSP through UART, a UART interrupt event will occur. The location and time information will then be abstracted from the raw serial data and saved if the GPS receiver has enough visible satellites.

- The DSP enters sleep mode when the navigation board is stationary. When a motion is detected by the PIC MCU, a logic high signal is sent to wake up the DSP core. The wake-up signal can also be generated by a push button on the printed circuit board (PCB).

7.3.2 System Power Management

In many applications, there are specific requirements to minimize power consumption and prolong the lifetime of the embedded system. C5505 has several means of managing the power consumption; the details are as follows:

- C5505 can selectively activate some subsystems while keeping other subsystems inactive by using software-controlled module clock gating;

- When not operating, the on-chip memory can be placed in a low-leakage-power mode while preserving the memory contents (DARAM/SARAM low-power modes);

- Independent power domains allow users to power down parts of the DSP to reduce static power consumption.

Since the navigation board uses batteries as its power supply, low power consumption is a primary requirement for software design. According to the signal sent from the motion detection MCU, DSP operates in either sleep mode or active mode and switches from one to the other alternatively to save power.

The power modes are described as:

Sleep Mode:

System clock is disabled; all the peripheral clocks are disabled.

On-chip memory in memory retention mode (placed in a low-leakage-power mode while preserving memory contents).

Disable the clock generator domain by placing the system clock generator in bypass mode and putting the phase-locked loop (PLL) in power-down mode.

Clear and disable all interrupts.

Enable the appropriate wake-up interrupt.

Active Mode:

> System clock and inter-integrated circuit (I2C), UART, *serial peripheral interface* (SPI) clocks are enabled.
>
> On-chip memory in active mode.
>
> Enable the clock generator domain.

Periodically check the general purpose input/output (GPIO) pin of motion detection MCU: If the output goes low, make DSP go to sleep mode again.

Similarly, CC2430 operates in its full functional mode PM0 (more information is provided in [16]) when the motion detection MCU indicates that the node is moving.

If the board is motionless for a certain time, CC2430 will enable an external I/O interrupt and enter PM3 with the lowest power consumption, in which all internal circuits that are powered from the voltage regulator are turned off. Once the board starts to move again, CC2430 will be woken up by the enabled external interrupt event sent from PIC MCU, enter PM0, and start from where it entered PM3 before.

7.3.3 DSP Memory Allocation

The navigation computation of C5505 has been tested by running the DSP in a code composer studio (CCS) emulator environment. Most of the variables involved in the computation are represented in the double precision format (64 bits). C5505 is a fixed-point DSP and the quantization errors yielded by truncating or rounding numbers during intensive computation cannot be overlooked. To study this, DSP outputs and the corresponding MATLAB® simulation results are compared (see Table 7.1).

The C5505 chip has three types of on-chip memory: 128 KB read-only memory (ROM), 256 KB single-access random access memory (SARAM), and 64 KB dual-access random access memory (DARAM). As the internal memory is sufficient for the navigation computations and system operations, the DSP requires no external RAM. The memory allocation map generated by CCS is shown in Table 7.2. This table shows that the total memory usage

TABLE 7.1

DSP Computation Accuracy Compared to MATLAB®

	Differences
Latitude	<0.000002° (0.21577 m)
Longitude	<0.000002° (0.21577 m)
North velocity	<0.4 m/s
East velocity	<0.4 m/s
Yaw	<0.020 rad/sec (1.1459°/sec)

TABLE 7.2

Memory Map Report Generated by CCS

Name	Origin (bytes)	Length (bytes)	Used (bytes, Hex)	Used (bytes, Dec)
MMR	00000000	000000c0	00000000	0
DARAM	000000c0	0000ff40	0000b98a	46.3848 K
SARAM	00030000	0001e000	0000a174	40.3633 K
SAROM_0	00fe0000	00008000	00000000	0
SAROM_1	00fe8000	00008000	00000000	0
SAROM_2	00ff0000	00008000	00000000	0
SAROM_3	00ff8000	00008000	00000000	0
			Bytes occupied in total	86.7481 K

is 86.7481 K bytes and 46.3848 K bytes are occupied in DARAM, including 35,000 bytes of heap, 5000 bytes of stack, and 5000 bytes of system stack.

7.3.4 ZigBee Node Software Design

A typical ZigBee network is a multihop network that enables low power consumption, low cost, and a low data rate for short-range wireless connections between battery-powered devices [21]. It is composed of three logical device types: coordinator, router, and end device. This section presents the software flow of sensor nodes and the sink node in ZigBee networks. The sink node works as coordinator, while the sensor nodes work as end devices.

7.3.4.1 Sink Node

SmartRF04EB with CC2430EM from TI [22] is used to develop and debug the sink node programs. It will initialize the board hardware configurations, create a network identifier, and broadcast previous access network (PAN) ID to start up the network. When the power is turned on, this coordinator will scan all channels and choose the one with the lowest energy level.

The task system is based on the multitask mechanism, and its main loop operates in the operating system abstraction layer (OSAL), which implements a cooperative, round-robin task-servicing loop. Each OSAL task has to be initialized first, including initialization of application object variables, instantiation of the corresponding application object(s), and registration with the applicable OSAL or hardware abstraction layer (HAL) system services. User-defined events are added to the task event processor to process all events for the task. The event processing flow of the sink node task event handler is shown in Figure 7.6.

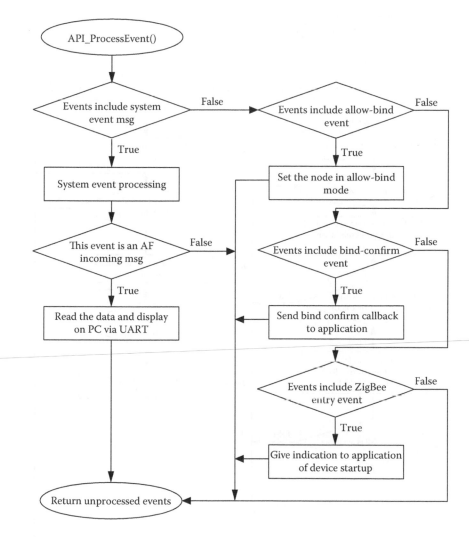

FIGURE 7.6
Event processing flow of the sink node task event handler.

7.3.4.2 Sensor Node

The sensor node works as an end device in the ZigBee network. This device can only receive and transmit information to the parent node (routers or the coordinator) and it has no routing capability. Therefore, it is a battery-powered node and it can sleep and wake up according to the network requirements. The navigation board with a CC2430 RF transceiver is used to work as an end device in this network. The event processing flow of its task event handler is shown in Figure 7.7.

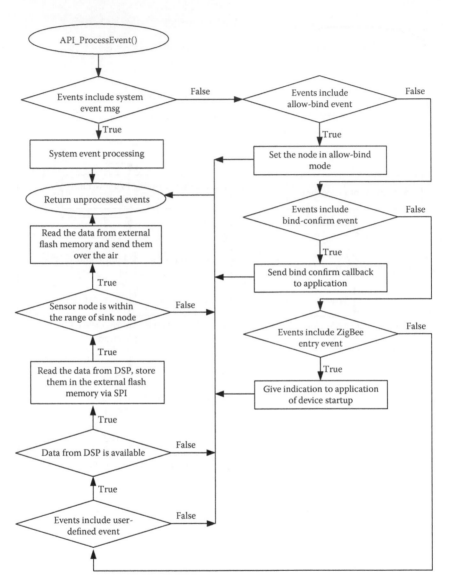

FIGURE 7.7
Event processing flow of the sensor node task event handler.

7.4 Test Results

7.4.1 Equipment and Setup

The MEMS inertial sensors being used in the laboratory calibration and field test I are the three-axis ±3 g accelerometer ADXL335 from Analog Devices

TABLE 7.3

Specifications for ADXL335 and LISY300AL

Parameter	Typical Value
Gyroscope LISY300AL	
Measurement range	±300 °/s
Sensitivity	3.3 mV/°/s
Drift	±2.4 °/s
Angular rate random walk	0.1 °/s/$\sqrt{Hz}$
Accelerometer ADXL335	
Measurement range	±3.6 g
Sensitivity	300 mV/g
Zero g bias stability	±10.8 mg
Acceleration random walk at X_{out}, Y_{out}	150 µg/$\sqrt{Hz}$ RMS
Acceleration random walk at Z_{out}	300 µg/$\sqrt{Hz}$ RMS

Inc. and the single-axis ±300°/s yaw rate gyroscope LISY300AL from ST Microelectronics. These accelerometers and gyroscopes operate on a voltage supply of 3.3 V. The sensitivity of the accelerometer is 300 mV/g and that of the gyroscope is 3.3 mV/°/s. The specifications for these analog inertial sensors are shown in Table 7.3.

The analog signals from the accelerometers and gyroscopes are sampled by a multichannel ADC that processes the inertial sensor measurements and converts them into digital signals. A 14 bit, 285 KSPS ADC from Analog Devices Inc. (AD7856) is used for sampling the inertial sensor measurements. A multiplexer is used for multichannel sampling at the input stage of ADC. Therefore, a time delay (Δt) exists between the sampling of two sequential channels [15]. Since the conversion time (Δt) is sufficiently short depending on the master clock signal (max. 6 MHz) for ADC, the effect of errors due to phase delays is negligible. The GPS receiver used in the field tests is LEA-5H from µ-blox.

Field test II was carried out to evaluate the performance of the navigation board. Three road trajectory tests were carried out in a land vehicle near the main campus of Queen's University, Kingston, Canada, on November 19, 2011.

In these tests, GPS LEA-5H was integrated with the navigation board, and inertial sensors ADIS16003, ADIS16060, and HMC6352 were used to provide the acceleration, the angular rate, and the heading angle respectively. Another GPS µ-blox EVK-5H was used as a reference to compare the positioning performance with the navigation board (see Table 7.4).

A laptop provided power for the GPS EVK-5H and CC2430EM-SmartRF04EB, which was used to receive the position data sent from the navigation board. The equipment setup is shown in Figure 7.8.

TABLE 7.4

Specifications for ADIS16003, ADIS16060, and HMC6352

Parameter	Typical Value
Gyroscope ADIS16060	
Measurement Range	±80 °/s
Sensitivity	0.0122 °/s/LSB
Drift	0.1 °/s
Angular Rate Random Walk	0.04 °/s/$\sqrt{Hz}$
Accelerometer ADIS16003	
Measurement Range	±1.7 g
Sensitivity	820 LSB/g
Zero g Bias Stability	±8.5 mg
Acceleration Random Walk at X_{out}, Y_{out}	110 µg/$\sqrt{Hz}$ RMS
Magnetic Sensor HMC6352	
Heading Accuracy	2.5 deg RMS
Heading Resolution	0.5 deg
Disturbing Field	min 20 Gauss
Max. Exposed Field	max 10,000 Gauss

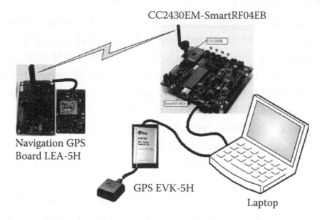

FIGURE 7.8
Equipment setup for field test II.

7.4.2 Real-Time Performance Analysis

The DSP real-time performance was evaluated by CCS C55xx Rev 3.0 cycle accurate simulator. Each assembly instruction is executed with a certain number of instruction cycles, and the duration of each mechanization computation and Kalman filtering computation can be calculated by counting the number of cycles that the program takes.

TABLE 7.5

Execution Time for the Program

	Number of CPU Cycles	Execution Time on DSP Operated with 100 MHz
Initialization		<2 ms
GPS Data Reading		90.909 ms (per second)
INS Data Reading		4.65 ms
Mechanization	84,709	0.8 ms
Kalman Filter	4,023,352	40 ms

As shown in Table 7.5, the number of instruction cycles required for one mechanization computation is 84,709; one Kalman filtering computation requires 4,023,352 cycles. When the DSP runs at 100 MHz, the duration of each instruction cycle is 10 ns. Therefore, the execution time for each step of mechanization computation takes 0.8 ms, while each step of Kalman filtering computation takes 40 ms. Based on these results, the duration of GPS and INS data reading can be calculated by running the prototype navigation board, recording the time, and using the CCS emulator to count how many times the navigation calculation loop repeats in this period. The duration of INS data reading for each time was approximately 4.65 ms on average, and the program took nearly 90.909 ms to read and parse the GPS data whenever it was updated. According to the software design introduced in Section 7.3, the navigation calculation is running in an infinite loop. In each step, the INS sensor reading, mechanization, and Kalman filtering computation will be executed successively. In contrast, GPS data reading runs in an interrupt service routine (ISR). Therefore, when a GPS update is available, the DSP with 100 MHz clock rate can run the navigation calculation loop 20 times and get 20 position results (as well as the velocity and heading information) per second; otherwise, if there is no GPS update, the DSP can get 22 position results per second.

7.4.3 Random Error Modeling

To derive the first-order noise variances and correlation times of MEMS gyroscope and accelerometer, an experimental dataset was collected from a two-hour static test to estimate the autocorrelation sequence (ACS) of inertial sensor errors. By removing the bias offsets from the raw measurements, the ACSs of two-axis accelerometer and gyroscope are calculated (Figure 7.9, Figure 7.10, Figure 7.11, and Table 7.6). These parameters are then used to model the stochastic errors of inertial sensors.

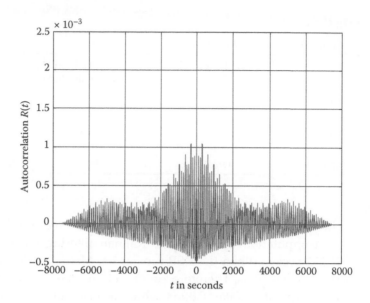

FIGURE 7.9
Autocorrelation of x axis accelerometer reading of two-hour stationary dataset after removing the bias offset (downsampled by a factor of 1000).

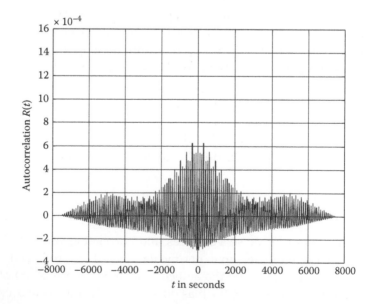

FIGURE 7.10
Autocorrelation of y axis accelerometer reading of two-hour stationary dataset after removing the bias offset (downsampled by a factor of 1000).

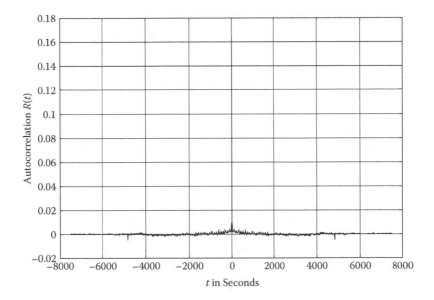

FIGURE 7.11
Autocorrelation of z axis gyroscope reading of two-hour stationary dataset after removing the bias offset (down sampled by a factor of 1000).

TABLE 7.6

First-Order Gaussian-Markov (GM) Model Parameters Used in Field Test I

	Acc.X	Acc.Y	Gyro.Z
Noise Variance σ^2	0.002062596	0.001597938	0.163237989
Correlation Time τ (sec.)	981.7	478	<100

7.4.4 Open Field Tests

7.4.4.1 Field Test I

The navigation system used in the first test is composed of ADXL335, LISY300AL, AD7856, GPS LEA-5H, and the DSP evaluation module TMS320VC5505EVM. Inertial sensors were mounted on a four-wheel cart. The GPS antenna was mounted on the top of this cart. Data acquisition was carried out using DSP to read both INS and GPS data. A laptop provided the power for the whole prototype board and recorded the data sent from the DSP.

This field test was carried out on September 23, 2010, along the race-track around the soccer field, on the main campus of Queen's University in Kingston, Ontario. The test field is an open area with a sufficient number of satellites available throughout the entire time period of the test. The average speed during this experiment was 1.45 m/s. Vibration of the inertial sensors

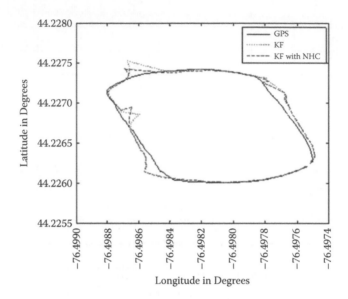

FIGURE 7.12
Field test I: GPS trajectory vs. KF trajectories with four simulated GPS outages.

was minimal due to the flatness of the ground. The test duration was 4 min 41 s. The total distance traveled was approximately 400 m.

In this test, nonholonomic constraints (NHC) were used to improve the accuracy of the loosely coupled GPS/INS integration system. Both trajectories obtained from the KF with and without the use of the NHC are depicted in Figure 7.12. To evaluate the performance of the navigation solution when GPS signals are absent, four 20-second GPS outages were simulated at different locations along the trajectory.

In addition, the position/velocity errors of the test results with and without NHC are given and compared in Table 7.7 and Table 7.8. It is clearly shown that the errors increased rapidly and with high amplitude during GPS outages.

The statistics of the position and velocity differences between integrated results and GPS outputs during four simulated GPS signal gaps are summarized in Table 7.7. The root mean square (RMS) values of the position errors are less than 14.5 m, while the maximum value of the position errors is 27.42 m. The RMS values of the velocity errors of these four GPS outages are not greater than 1.44 m/s.

To assess the accuracy improvement, Table 7.8 compiles the corresponding differences between the position/velocity error statistics provided by the Kalman filter with and without the use of NHC.

The dataset in this table shows that both velocity and position estimations of KF with NHC give better results in comparison with the KF algorithm without the use of NHC, although in some cases they provide a comparable level of accuracy. This result confirms that the NHC guarantees an improved performance during short-term GPS outages.

TABLE 7.7

Position/Velocity Errors during GPS Outages

	Outage 1		Outage 2		Outage 3		Outage 4	
Northing Position Errors	Mean	8.25	Mean	1.89	Mean	3.96	Mean	11.56
in Meters	RMS	9.83	RMS	2.23	RMS	5.58	RMS	14.28
	Max.	17.44	Max.	4.61	Max.	12.81	Max.	26.97
Easting Position Errors	Mean	1.81	Mean	3.92	Mean	6.82	Mean	1.23
in Meters	RMS	2.23	RMS	5.03	RMS	9.12	RMS	1.76
	Max.	3.98	Max.	10.21	Max.	19.04	Max.	4.94
Position Errors in Meters	Mean	8.49	Mean	4.40	Mean	7.98	Mean	11.65
	RMS	10.08	RMS	5.50	RMS	10.69	RMS	14.39
	Max.	17.89	Max.	11.21	Max.	22.95	Max.	27.42
Northing Velocity	Mean	0.78	Mean	0.32	Mean	0.70	Mean	1.37
Errors in M/sec	RMS	0.83	RMS	0.38	RMS	0.85	RMS	1.44
	Max.	1.36	Max.	1.08	Max.	1.51	Max.	1.97
Easting Velocity Errors	Mean	0.17	Mean	0.61	Mean	0.97	Mean	0.41
in M/sec	RMS	0.19	RMS	0.64	RMS	1.09	RMS	0.55
	Max.	0.30	Max.	0.93	Max.	1.67	Max.	1.11

TABLE 7.8

Accuracy Improvement by NHC

	Outage 1		Outage 2		Outage 3		Outage 4	
Latitude Error in	Mean	0.05311	Mean	−0.15141	Mean	−2.73988	Mean	−2.71550
Meters	RMS	0.07152	RMS	0.25123	RMS	−4.27284	RMS	−3.29470
	Max.	0.16383	Max.	0.37841	Max.	−10.46811	Max.	−6.41486
Longitude Error	Mean	0.06345	Mean	−1.43684	Mean	0.35443	Mean	−0.02630
in Meters	RMS	0.08462	RMS	−1.84654	RMS	0.51679	RMS	−0.08213
	Max.	0.18479	Max.	−3.63789	Max.	1.29497	Max.	−0.16886
Position Error in	Mean	0.06614	Mean	−1.31906	Mean	−0.44483	Mean	−2.68400
Meters	RMS	0.08871	RMS	−1.46655	RMS	−0.96712	RMS	−3.27541
	Max.	0.20141	Max.	−2.95220	Max.	−2.47791	Max.	−6.31721
North Velocity	Mean	0.00881	Mean	0.02552	Mean	−0.48160	Mean	−0.30842
Error in M/sec	RMS	0.00960	RMS	0.06688	RMS	−0.55773	RMS	−0.33708
	Max.	0.01426	Max.	0.09853	Max.	−0.92419	Max.	−0.52490
East Velocity	Mean	0.00020	Mean	−0.18625	Mean	0.06667	Mean	−0.00213
Error in M/Sec	RMS	0.00923	RMS	−0.20072	RMS	0.08750	RMS	0.01225
	Max.	0.01536	Max.	−0.28659	Max.	0.20073	Max.	0.04880

7.4.4.2 Field Test II

Three trajectories obtained from the GPS EVK-5H and the navigation for an in-vehicle scenario are depicted in Figure 7.13, Figure 7.14, and Figure 7.15. Several GPS outages took place at different locations along the trajectory,

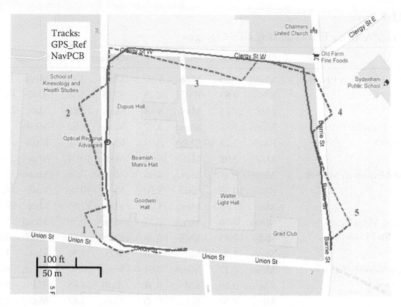

FIGURE 7.13
Field test II (A): GPS trajectory vs. in-vehicle trajectory with five GPS outages.

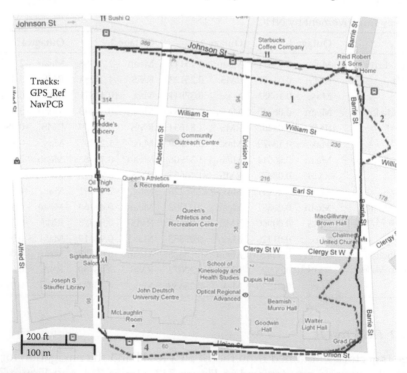

FIGURE 7.14
Field test II (B): GPS trajectory vs. in-vehicle trajectory with four GPS outages.

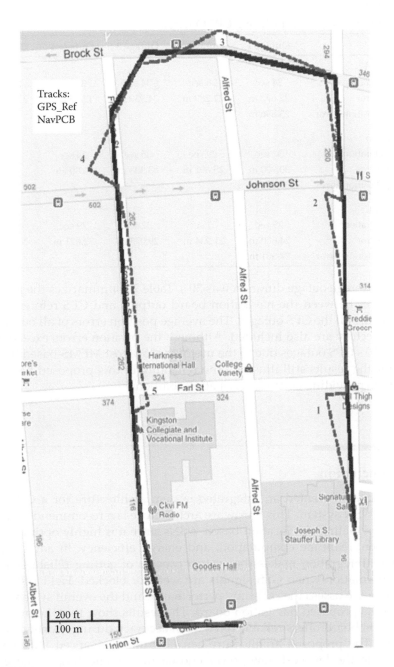

FIGURE 7.15
Field Test II (C): GPS trajectory vs. in-vehicle trajectory with five GPS outages.

TABLE 7.9

Field Test II: Position Errors during GPS Outages

	Outage 1	Outage 2	Outage 3	Outage 4	Outage 5
Trajectory A					
Outage Duration	24 sec	28 sec	34 sec	36 sec	30 sec
Position Error	22.672 m	22.289 m	34.939 m	17.453 m	21.127 m
Average Position Error	23.696 m				
Trajectory B					
Outage Duration	32 sec	29 sec	35 sec	34 sec	
Position Error	30.900 m	25.692 m	33.308 m	26.803 m	
Average Position Error	29.176 m				
Trajectory C					
Outage Duration	29 sec	32 sec	32 sec	29 sec	33 sec
Position Error	24.600 m	21.934 m	29.073 m	32.021 m	28.879 m
Average Position Error	27.301 m				

and the average outage duration was 30 s. Table 7.9 summarizes the position differences between the navigation board outputs and GPS reference outputs during all the GPS outages. The average position errors of all outages in each trajectory are also included. Although the position errors exceed 20 m during 30 s GPS outages due to the use of the low-cost MEMS-based inertial sensors, the results still illustrate the reliability of this proposed hardware system in providing accurate navigation solution.

7.5 Conclusion

This chapter presented an integrated system architecture for a GPS/INS-enabled WSN system. This hardware architecture is a recommended implementation method for location-based WSNs since it is highly optimized for dimension, real-time computation, and energy efficiency. In addition, the INS/GPS integration makes the system capable of getting reliable navigation information when GPS signals are weak or blocked. Field tests were carried out to assess the real-time performance and the overall system accuracy of the proposed navigation board. The results show that the proposed navigation board offers reliable navigation accuracy during short-term GPS outages. The proposed architecture can host more advanced INS devices (instead of the low-cost, low-power chosen in this chapter), which could bring more accuracy to the system during outages but at probably higher cost and higher energy consumption. Current trends in the field show that new generations of INS devices are getting even smaller in size and weight and better in terms of energy efficiency. Finally, the chapter has shown that wireless sensor networks are realistically able to provide the GPS/INS location information of the nodes, which enables them to be used for cognitive routing and other location-based cognitive protocols.

References

1. Y.B. Ko and N.H. Vaidya, "Location-aided routing (LAR) in mobile ad hoc networks," *Wireless Networks* 6, no. 4, 2000, pp. 307–321.
2. M. Zorzi and R.R. Rao, "Geographic random forwarding (GeRaF) for ad hoc and sensor networks: Multihop performance," *IEEE* Transactions on Mobile Computing 2, no. 4, 2003, pp. 337–348
3. V. Hnatyshin, "Design and Implementation of an OPNET model for simulating GeoAODV MANET routing protocol," in *Proceedings of the OPNETWORK 2010 International Conference*, Washington, D.C., 2010.
4. D. Dipankar and S.B.R. Nabendu Chaki, "LACBER: A new location aided routing protocol for GPS scarce MANET," *International Journal of Wireless & Mobile Networks (IJWMN)* 1, August 2009.
5. G. Zhang, "A low cost integrated INS/GPS system," *UCGE Report,* no. 20078, 1995.
6. G.H. Elkaim, M. Lizarraga, and L. Pederseny, "Comparison of low-cost GPS/INS sensors for Autonomous Vehicle applications," *IEEE/ION Conference*, 2008, pp. 1133–1144.
7. R.T. Kelley, I.N. Katz, and C.A. Bedoya, "Design, development and evaluation of an Ada coded INS/GPS open loop Kalman filter," in *Proceedings of the IEEE National, Aerospace and Electronics Conference*, 1990, pp. 382–388.
8. A. Chatfield, *Fundamentals of High Accuracy Inertial Navigation*, American Institute of Aeronautics and Astronautics, 1997.
9. R.M. Rogers, *Applied Mathematics in Integrated Navigation Systems*, 3rd ed. AIAA education series, Reston, Va.: American Institute of Aeronautics and Astronautics, 2007.
10. A. Noureldin, A. Osman, and N. El-Sheimy, "A neuro-wavelet method for multi-sensor system integration for vehicular navigation," *Measurement Science & Technology* 15, no. 2, 2004, pp. 404–412.
11. K.W. Chiang, A. Noureldin, and N. El-Sheimy, "A new weight updating method for INS/GPS integration architectures based on neural networks," *Measurement Science & Technology* 15, no. 10, 2004, pp. 2053–2061.
12. E.H. Shin, *Accuracy Improvement of Low Cost INS/GPS for Land Applications*, internal report, Department of Geomatics Engineering, University of Calgary, Canada, Report #20156, 2001.
13. J. Shang, G. Mao, and Q.T. Gu, "Design and implementation of MIMU/GPS integrated navigation systems," *IEEE Position Location and Navigation Symposium*, 2002, pp. 99–105.

14. P.P. Qu, L. Fu, and X. Zhao, "Design of inertial navigation system based on micromechanical gyroscope and accelerometer," *CCDC 2009: 21st Chinese Control and Decision Conference, Vols. 1–6, Proceedings*, 2009, pp. 1351–1354.

15. V. Agarwal, H. Arya, and S. Bhaktavatsala, "Design and development of a real-time DSP and FPGA-based integrated GPS-INS system for compact and low power applications," *IEEE Transactions on Aerospace and Electronic Systems* 45, no. 2, 2009, pp. 443–454.

16. ADS8364, *Texas Instruments, datasheet*. Available from: http://focus.ti.com/lit/ds/symlink/ads8364.pdf.

17. ADIS16060, *Analog Devices Inc., datasheet*. Available from: http://www.analog.com/static/imported-files/data_sheets/ADIS16060.pdf.

18. ADIS16003, *Analog Devices Inc., datasheet*. Available from: http://www.analog.com/static/imported-files/data_sheets/ADIS16003.pdf.

19. LEA 5H, *u-blox, datasheet*. Available from: http://www.u-blox.com/images/downloads/Product_Docs/LEA-5x_DataSheet_%28GPS.G5-MS5-07026%29.pdf.

20. CC2430, *Texas Instruments, datasheet*. Available from: http://focus.ti.com/general/docs/lit/getliterature.tsp?genericPartNumber=cc2430&fileType=pdf.

21. *Z-Stack Developer's Guide,* Texas Instruments, Document Number: F8W-2006-0022.

22. *CC2430DK Development Kit User Manual,* Texas Instruments. Available from: http://focus.ti.com/lit/ug/swru133/swru133.pdf.

23. C. Tang, *A Hardware Architecture for GPS/INS-Enabled Wireless Sensor Networks*, MSc thesis, Queen's University, 2012.

24. F. Aalamifar, *Design and Hardware Implementation of a Cognitive Wireless Sensor Network: Application to Environment Monitoring,* MSc thesis, Queen's University, 2011.

25. F. Aalamifar, G. Vijay, P.A. Khozani, and M. Ibnkahla, "Cognitive wireless sensor networks for highway safety," *NSERC DIVA Workshop, ACM Proceedings,* Ottawa, September 2011.

Index